Forschungs-/Entwicklungs-/ Innovations-Management

Herausgegeben von
H. D. Bürgel (em.), Stuttgart, Deutschland
D. Grosse, Freiberg, Deutschland
C. Herstatt, Hamburg, Deutschland
H. Koller, Hamburg, Deutschland
M. G. Möhrle, Bremen, Deutschland

W0079794

Die Reihe stellt aus integrierter Sicht von Betriebswirtschaft und Technik Arbeitsergebnisse auf den Gebieten Forschung, Entwicklung und Innovation vor. Die einzelnen Beiträge sollen dem wissenschaftlichen Fortschritt dienen und die Forderungen der Praxis auf Umsetzbarkeit erfüllen.

Herausgegeben von

Professor Dr. Hans Dietmar Bürgel (em.), Universität Stuttgart

Professor Dr. Hans Koller Universität der Bundeswehr Hamburg

Professorin Dr. Diana Grosse, vorm. de Pay, Technische Universität Bergakademie Freiberg

Professor Dr. Martin G. Möhrle Universität Bremen

Professor Dr. Cornelius Herstatt Technische Universität Hamburg-Harburg

Jan Bierwald

Specialization in Online Innovation Communities

Understand and Manage Specialized Members

With a foreword by Univ. Prof. Dr. Cornelius Herstatt

Jan Bierwald
Hamburg, Germany

Dissertation Technischen Universität Hamburg-Harburg, 2013

ISBN 978-3-658-05317-8   ISBN 978-3-658-05318-5 (eBook)
DOI 10.1007/978-3-658-05318-5

The Deutsche Nationalbibliothek lists this publication in the Deutsche Nationalbibliografie; detailed bibliographic data are available in the Internet at http://dnb.d-nb.de.

Library of Congress Control Number: 2014933591

Springer Gabler
© Springer Fachmedien Wiesbaden 2014

Printed on acid-free paper

Springer Gabler is a brand of Springer DE.
Springer DE is part of Springer Science+Business Media.
www.springer-gabler.de

Foreword

Online Innovation Communities (OIC), virtual groups of often more than thousands of users, are gaining more attention as drivers for product and service innovations. Nowadays many firms are already cooperating closely with these OIC within the innovation process.

The work of Mr. Bierwald investigates how members of Online Innovation Communities (OICs) focus their contributions and align their behavior to altering circumstances in order to understand the division of labor meaning the degree and direction of the individual's specialization within these OIC. OIC are a rather new organizational form, which is characterized by the goal to innovate and the voluntary participation and contribution of their members. Therefore the division of labor cannot be assigned hierarchically and has to emerge from the concurrence of different choice rationales. The specialization by one's own choice in communities of volunteers is a new and theoretically relevant research topic. The current state of research is lacking this phenomenon; hence a comprehensive understanding of this member behavior is essential.

Thus far academic empirical research is mainly focused on the individual participation motivation and the community organization. Insights about the micro-dynamic within these networks are only rudimentary. The dissertation by Mr. Bierwald contributes to a deeper understanding of these mechanisms and addresses three core research questions: 1) How do community members focus their contributions? 2) How does this specialization affect other behavioral patterns such as activity and community joining? 3) How does the community management ensure the permanent transfer of special knowledge into the community?

To study the individual contribution behavior of each community member, Mr. Bierwald conducted a content analysis of 7.362 mails sent by 105 members over a time period of 62 months. Additionally he applied co-citation, time series and social network analyses of the community. In order to support his hypotheses for the member behavior he uses linear regressions models and non-parametric tests. This dissertation provides several novel and interesting findings: Members focus their contributions based on their familiarity with the development object; minimizing their participation costs is the main driver behind the behavior of community members, especially strongly specialized members; core members of the community have the ability to transfer knowledge permanently into the organization; entrepreneurial members are valuable contributors but require intensified supervision.

In this work Mr. Bierwald provides new, additional insights regarding the individual behavior of community members, especially the entrepreneurial spirit as driver for certain behavioral patterns. He derives clear measures for the community management how to improve the recruiting of new members and the controlling as well as the retaining of existing members. Mr. Bierwald pursues a courageous and autonomous approach. He derives new, for research and management relevant research hypotheses and supports several of them with empirical findings. In cases where empirical findings could not support the derived hypotheses he seeks out for alternative approaches and provides additional theoretical and practical enriching findings. I believe that the findings of this work are relevant for researchers and practitioners and lead to differentiated insights of the innovation in communities. Altogether, I consider this present dissertation as a very readable contribution to today's research.

Univ. Prof. Dr. Cornelius Herstatt
Hamburg, January 2014

Acknowledgements

My dissertation was a long and quite exhausting journey which I could not have successfully completed without the support of many people.

First, I would like to thank my doctoral supervisor Prof. Dr. Herstatt for giving me the opportunity to write my PhD thesis at the Institute for Technology and Innovation Management at the Hamburg University of Technology. I admire your patience, enthusiasm, permanent support and openness to a variety of research methods and approaches, which I know cannot be taken for granted. I will always appreciate the support of my academic mentor, Prof. Dr. Christina Raasch, also called Tina. We had so many intense but also very fruitful discussions and we learned so much from each other. You taught me everything I know about research and hopefully I was a perfect case example for guiding an enthusiastic but hardly teachable PhD candidate.

I met many research fellows at the institute during my four years but five of them need to be emphasized. Viktor for lending me some of "his" research methods and our discussions about the greatest sport in the world. Daniel for our shared interest in real estate and his critical (sometimes annoying) but always helpful questions that increased the quality of my thesis tremendously. Tim for his valuable advices and endorsements in any professional and unprofessional situation. Niclas for thousands of laughs and finally Sarah my best female friend for your support over the last six years.

Additionally, I could not have accomplished all this without the trust and supportiveness of the Premium Community with honorably mentioning Uwe Lübbermann. You let me into the core of your community without any restrictions and gave me valuable input whenever we talked.

Anyway, with my last words I would like to express my greatest gratitude to my former girlfriend Tanja, who became my wife during my PhD studies and gave birth to our little miracle Luca. I will never forget your support, especially in the most debilitating parts of this journey.

List of Contents

List of Tables

List of Figures

List of Abbreviations

Cf.	compare
E.g.	Exempli gratia
Et al.	Et alii
GAC	Group activity change
MAY	Member activity change
MOI	Man/ Men-on-the-inside
No.	Number
NOF	New Organizational Forms
OC	Online Community
OIC	Online Innovation Community
OSI	Open Source Innovation
OSS	Open Source Software
RCT	Rational Choice Theory
SNA	Social network analysis

1 Introduction

The array of organizational forms is not constant. Changing environmental conditions (Stinchcombe 2000) frequently force organizations to realign their structure and processes to their purposes (Miles et al. 1978). New organizational forms (NOF) arise. Over the last two decades technological breakthroughs in computer-mediated communication have caused major environmental changes. Therefore, the internet noticeably serves as a breeding ground for various NOF. One of the best known NOF is the online community in which a group of volunteers exchange various forms of content electronically (Rheingold 2000). Online innovation communities (OIC) are a more specific form of this NOF. In these organizations sometimes thousands of members contribute voluntarily to a jointly developed innovative outcome (Raasch et al. 2009). This outcome can be manifold, e.g. software, physical products or content. In many cases users of the outcome represent the majority of the involved community members. Hence, collaborative innovating in virtual organizations is obviously closely related to the phenomenon of user innovation. Users innovate solely to benefit from their development and therefore they often develop fundamentally different solutions with high novelty (von Hippel 2006). Even though the phenomenon of user innovation has been long known (von Hippel 1976), it has gained significant momentum since the rise of the internet. It significantly improved the knowledge acquisition for users and enabled them to innovate in areas formerly unthinkable. Additionally the internet changed the way of communicating tremendously facilitating the joint development by globally dispersed users.

The success of these OIC e.g. Linux or Wikipedia jeopardizes the market shares of profit oriented organizations. Chandler et al. (1962) already stated the essentiality of innovating for a firm's competitive advantage[1] in 1962. Nowadays, academics (e.g. Furman et al. 2002) and practitioners (e.g. Jaruzelski and Dehoff 2008) are still not tired of emphasizing the importance of innovating for a firm's success. Instead of striving against or ignoring OIC firms have started to treat the communities and their members as valuable resources. Analyses show that the majority of firms[2] have already established the instrument of OIC to benefit from user contributions during the innovation process (Mahr and Lievens 2012). The firms have more and more shifted from a closed to an open innovation model. In the new model, firms use external and internal ideas as well as external and internal paths for bringing them to the market (Chesbrough 2004). They increase the permeability of their organizational boundaries to utilize resources outside of the firm. By opening up their innovation processes

---

[1] A firm holds a competitive advantage "when it is implementing a value creating strategy not simultaneously being implemented by any current or potential competitor" (Barney 1991, p. 102).
[2] E.g. 80% of the firms listed in the S&P 500 are engaged in OIC (Mahr and Lievens 2012).

they gain the advantage of acquiring and leveraging external knowledge. Nevertheless they also put valuable technological knowledge at risk of acquisition by partners outside of the firm (Oxley and Sampson 2004). Thus, understanding the organization of OIC and the micro dynamics within these networks are key challenges for today's innovation success (Mahr, Lievens 2012).

1.1 Research gap

Modularity is one of the major organizational concepts shaping the design of OIC and influencing the micro dynamics within these networks. The importance and benefits of modularity for OIC have been well researched. Especially technological modularization of the innovative outcome with the coexistence of several modules has been observed in almost all larger OIC (de Laat 2007). As "organizations should be designed to reflect the nature of the tasks that they perform" (MacCormack et al. 2012, p. 2), the organizational form of OIC should also be characterized by a high modularity level (MacCormack et al. 2006). Most of the benefits from modularity have been proven as substantially favorable for OIC. The decrease in communication costs for example limits free-riding and lurking[3] of community members (e.g. Baldwin and Clark 2006). Furthermore modularity, based on the economies of substitutions concept, facilitates the development process. Modules or at least parts of it can be easily transferred between different projects (e.g. Haefliger et al. 2008). Thus, one of the major benefits has gained almost no attention by OIC research: the division of labor leading to specialization. Smith already argued in the 18th century, that division of labor and specialization of workers are crucial for the wealth of nations (Smith 1937). Is this basic organizational concept not crucial for OIC? Despite the importance of modularity for most OIC could its major benefit does not affect the community at all? Are thousands of volunteers able to jointly develop an innovative outcome without carefully focusing their contributions? That is hard to conceive and initial work from research scholars indicates the appearance and importance of specialization in OIC.

The exploratory work of von Krogh et al. (2003) unveils different degrees of specialization among members. The small group of core members[4] shows a high degree of generalization, meaning that the majority of community members tend to specialize. They complement this finding by showing that newcomers joining the community strongly specialize on certain modules. Furthermore, Nambisan and Baron (2010) identifies a focus on certain topics by

---

[3] Free-riding and lurking are detailed later in this thesis.
[4] A detailed explanation of different member groups of OIC follows later in this thesis.

different community members. They analyze if members tend to contribute to the community or to the firm sponsoring the community. Thus far, researchers argue solely with already available knowledge as specialization driver (von Hippel and von Krogh 2003). Nevertheless, these findings need to be complemented and extended. Simply unveiling the occurrence of specialization and discussing one rationale behind this behavior are not sufficient to comprehensively understand member specialization in OIC. Therefore, in this thesis, factors beyond the member's knowledge level influencing specialization should be identified. Afterwards the research gap is expected to be filled by addressing three research questions. First, how do community members focus their contributions? Second, how does an altering community focus influence the contribution activity of specialized members? Third, how is knowledge from specialized individuals integrated permanently into the community?

1.2   Research approach and contributions

Considering the type of research questions defined (Yin 2009) and the objective to understand the micro dynamics in OIC (Eisenhardt 1989) a case study approach is chosen. The contribution behavior of community members needs to be examined in a micro-analytical way (Kuk 2006). To identify the contribution focus of individual members a detailed content analysis of mails or posts is required. As this process is very time-consuming a single case study is selected to not exceed the resources available to one researcher (Yin 2009). Therefore the raw data, the mailing list of the 'Premium' community that jointly develops and improves several consumer goods, is analyzed in depth by the author of this study. This comprises analyzing the content of approx. 7,400 mails from a 5-year period and adapting the co-citation analysis to determine the communication intensity between the individual community members. The individual community members are grouped into four different groups based on their familiarity with the community outcome, e.g. private and commercial members. Subsequently several statistical assessment are conducted to support the previously derived hypotheses for each of the three research questions. These research hypotheses were derived prior starting with the empirical study by applying different theories of individual behavior, e.g. social cognitive and choice theory. The results from the quantitative analyses show additional interesting insights about the specialization behavior of the individual member groups. Therefore, the prepared data is additionally analyzed using qualitative methods, such as social network and homogeneity analysis, to detail and enrich the findings.

The findings of this study provide valuable insights for today's research and management. Besides the study of Nambisan and Baron, this study is the only empirical quantitative work studying member specialization in OIC and outranges their sample size by almost 500 %[5]. Furthermore the content provided by the different community members are coded on a considerably higher detail level[6]. This enables a clear differentiation of community member groups regarding their contribution focus for the first time. Additionally it is the first longitudinal study referring to specialization as it covers a time frame of 62 months. Therefore, the impact of changes in the community communication on the behavior of specialized members is initially studied. By applying the co-citation analysis in an innovative manner the process of transferring special knowledge permanently into the community via certain individuals is observed. All these enrichments for the literature about community member behavior directly help the community management to increase project success. The community management can align their recruiting, controlling and organizational measures knowing: Exactly which type of member contributes which content; how specialized members behave confronted with changes in the community; how to empower and motivate individuals to hold special knowledge permanently in the community.

1.3 Structure of dissertation

In addition to the general introduction into the research objectives in *Chapter 1*, this thesis consists of three main parts that are sub-divided into two to three chapters each:

Part A provides the reader with the phenomenological and theoretical background required for a comprehensive understanding of member specialization in OIC as well as the detailed research setting.

Therefore, *Chapter 2* outlines different concepts of organizational behavior. This includes the arising of new organizational forms, online communities, modularity and the division of labor. Furthermore, different categories of member behavior in OIC are defined after user innovation and community-based innovation have been delineated as major research streams from innovation research. The second chapter closes with identifying member specialization as research gap. Subsequently three research questions are defined, which are addressed in this thesis to fill this gap.

Chapter 3 details the rational choice and social cognitive theory as well as key individuals in the innovation process, such as boundary spanners and promotors. These theories are well suited to predict, understand and explain member behavior in OIC. Finally, for each of the

---

[5] 7,362 mails vs. 1,339 posts
[6] Seven categories vs. two categories

three research questions a main hypothesis is derived. This is based on the theoretical background and empirical findings from various research scholars to related topics.

Part B describes the empirical study conducted to answer the research questions formulated in the first part.

Chapter 4 starts with defining the case study approach as the appropriated research method. The selected OIC 'Premium', in which various consumer products are collaboratively developed and improved by more than 100 active community members, is described in detail. To increase the evidence of the results the selected case represents a typical case. For verification 'Premium' and its members are linked to generally accepted frameworks of OIC characteristics. To unveil the behavior of its members a detailed content analysis is conducted. Therefore, approx. 7,400 mails from a 5-year period are coded to seven different categories. Additionally a co-citation analysis and its adaption to community communication data is explained and conducted.

In *Chapter 5,* the previously derived three main research hypotheses are operationalized by transferring them into 17 detailed hypotheses. These are subsequently statistically assessed with non-parametrical tests and linear regression models. The chapter closes with a brief discussion of the results from the quantitative analyses and the reasoning for further investigations regarding the researched phenomenon.

Chapter 6 shows results from several qualitative methods, such as social network analysis and semi-structured interviews with certain community members. The reason for conducting additional analyses is to find initial evidence for the newly formulated proposition based on the results from the first analysis part. Additionally some explanatory elements for the unpredicted behavioral patterns of the individual community members are identified.

Part C summarizes the results from the empirical study, discusses the findings by relating them to empirical studies others and outlines their impact on research and management.

Chapter 7 synthesizes the results from the quantitative and qualitative analyses from the fifth and sixth chapter for the different community member groups.

In the final *chapter 8,* four different aspects are considered. Starting with an extensive and critical discussion of all findings from the empirical study, contributions made to research are discussed. The chapter continues by outlining the limitations of this research work as well as suggestions for further research in this field. Finally concrete management measures to improve community success are derived.

Part A: Foundation of dissertation

Part A of the thesis lays the foundation to successfully investigate the previously outlined research objectives. Even though member specialization in OIC seems to be strongly anchored within innovation research, many other research streams can provide substantial support for a comprehensive exploration of this phenomenon. As mentioned previously, especially organizational instruments and functions are of paramount importance for the way OIC work and individual community members behave. Therefore, different organizational functions and instruments are detailed in order to describe and understand the concept of online communities and the way labor is effectively organized to enable and optimize specialization. Starting with users as the locus of innovation, the way individual community members of OIC behave while contributing to innovative outcomes is outlined to set the appropriated framework for community member specialization. As the aim of this study is to predict, unveil and explain the behavior of individual community members, theories of individual behavior are predestined to build the theoretical foundation (Davis and Luthans 1980). This chapter is of paramount importance. First, by thoroughly reviewing current literature of member behavior in OIC and the underlying organizational concepts the research questions are identified. Second, the factors influencing this behavior and the selected theories of individual behavior help to derive the research hypotheses. The following figure visualizes the concepts, theories and phenomena detailed while proceeding:

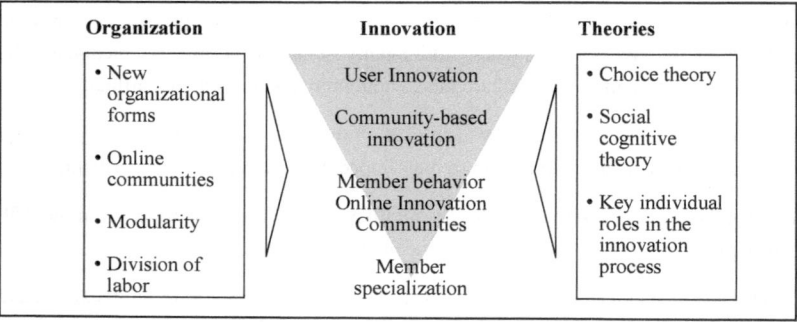

Figure 1: Phenomenological and theoretical foundation

2 Phenomenological background

This chapter outlines essential concepts concerning organization and the actual core phenomenon of this thesis 'member specialization in OIC' is exposed thoroughly and embedded into Innovation research. This directly points to the research gap, which is expected to be filled by addressing three research questions.

2.1 Organization

Organization is a widely used term in today's research and management. Commonly organization is divided into three categories: Institutional, instrumental and functional organization (Schulte-Zurhausen 2005). Referring to the institutional view, an organization is a social system, clearly separated from its environment, with a clear purpose and a formal structure (Cushway and Lodge 1999). The instrumental view of organization looks at the formally or informally defined rules between elements (e.g. people, information, tasks) required to establish and maintain the formal structure, respectively the organizational form, of a social system (Schulte-Zurhausen 2005). The third view describes the management function of organization, meaning all activities related to the planning, implementation and enforcement of organizational rules (Schulte-Zurhausen 2005). Considering the research scope of this thesis, to investigate how to manage member specialization in OIC, especially the instrumental and functional view on organization is of utmost importance. Therefore, before referring to various forms of online communities (OC), the arising of new organizational forms in general is explained. Furthermore, one major organizational function, modularity, is essential for OC on the one hand and for the concept of specialization on the other hand. Therefore, this function is detailed before the concept of labor division and specialization is defined at the end of this chapter.

2.1.1 New organizational forms

"The concept of organizational form refers to those characteristics of an organization that identify it as a distinct entity and, at the same time, classify it as a member of a group of similar organizations" (Romanelli 1991, p. 82). Organizational forms have always been a well-studied research objective and various basic forms have been identified and thoroughly analyzed (Burton et al. 1998). Well-known basic organizational forms are for example the divisional, the functional and the matrix form. These forms "can roughly be described by analyzing [the] methods of grouping activities, the number of hierarchical levels, and by the extent to which management is divided into various functional areas" (van den Bosch et al.

1999, p. 555). Considering these three differentiation criteria and by referring to Volberda (1999), the three mentioned basic organizational forms can be described as following: When using the divisional form, the organizational activities are grouped by combinations of products and markets resulting in various divisions. Therefore, only a few levels of hierarchies are installed and the functional division of the management is limited. In the functional form, the activities are grouped under functional managers leading to several hierarchical levels and the functionality level of the management varies. In the matrix form activities are grouped by functional departments and autonomous organizational units. The hierarchical levels are limited but duplicated due to the dual grouping of the activities and the management tasks are heavily functionalized.

The array of organizational forms is not constant. By "rearranging their structure of roles and relationships and their managerial processes" (Miles et al. 1978, p. 547) organizations refine their mechanism to achieve their purposes. Therefore, new organizational forms (NOF) arise frequently and the organizational forms known nowadays are a result of innovative organizational actions as a response to changing environmental conditions in the past (Stinchcombe 2000). Within the last two decades, the environmental landscape for organizations has changed tremendously. Globalization, knowledge-based competition and increased volatility forced organizations to rethink their organizational design (Daft and Lewin 1993) and served as breeding ground for the arising of NOF. In order to cope with the increased importance of human capital (Romanelli 1991) and knowledge (van den Bosch et al. 1999) as well as the increased permeability of the boundaries between organizations and markets (Foss 2002), organization had to undergo several organizational design changes resulting in many NOF. These NOF are characterized by e.g. more temporary constellations of people, increased lateral communication within the organization, intensified outsourcing and a higher level of organizational modularity[7] (Hedlund 1994). Even though the radical environmental changes in the last decades issued enormous challenges to organizations, in parallel technological breakthroughs in manufacturing and especially in computer-mediated communication enabled various NOF (Daft and Lewin 1993). One of these NOF, that offers organizations new ways to absorb knowledge and organize labor, is the virtual or online community that is detailed in the following chapter.

---

[7] The term modularity in general and organizational modularity in particular are described in detail in the chapter 'Modularity'

2.1.2 Online communities

A community is a private, exclusive group of several individuals separated fom the society normally sharing a common geographical location (Tönnies 2012). Prior to the rise of the internet these communities were mostly face-to-face communities characterized by physical meetings of the community members and a simultaneous interaction between them. Over the recent two decades, facilitated by the internet the new organizational form of virtual or online community has emerged strongly. Within these communities, the individual community members are geographically dispersed, mainly interacting via internet (Blanchard and Horan 1998). Nevertheless, physical meetings between community members can occur, but they do not represent the main intention to participate in OC. In recent literature, numerous definitions of OC exist covering different characteristics of this concept. Rheingold (2000) investigating virtual communities as one of the first researchers states, "a virtual community is a group of people [...] who exchange words and ideas through the mediation of computer bulletin boards and networks" (Rheingold 2000, p. 58). With further research on OC, shared interests and trust among the members are identified as important characteristics. OC "can be defined as affiliate groups whose online interactions are based upon shared enthusiasm for, and knowledge of, a specific consumption activity or related group of activities" (Kozinets 1999, p. 254). "People come to virtual communities to exchange information - either by providing it to others or by soliciting it from others. This exchange is based upon the trust the members have in each other [...]" (Ridings et al. 2002, p. 288). Summarizing these and other definitions the following characteristics describe the basic concept of OC thoroughly:

- A group of volunteers who exchange various forms of content and communicate.
- The group mainly communicates using technical platforms such as blogs, forums, e-mail, chat or instant messaging.
- The individual group members share a common goal, identity or interest.
- Trust and expected reciprocity among the community members enables the observing of shared values, rules and norms and helps to establish some kind of community culture (Tietz 2007).

Diverse dimensions exist to classify OC. The most granular typology is presumably delivered by Dube et al. (2006). Their proposal to differentiate OC includes four main dimensions: demographics, organization, membership characteristics and technological environment. Demographics include for example the level of maturity, the age and the basic orientation of the community. The design of the creation process, the leadership model and the degree of integration of the OC into other organizations are some of the sub-dimensions covered by organization. They characterize the community members according to several sub-dimensions, such as group size, geographic dispersion, prior community experience, and

9

cultural diversity. The technological environment is defined by the availability of certain information and communication technologies and the degree of relying on those. The initiator of the OC is an additional dimension that exceeds the framework of Dube et al. (2006). OC are either founded by community members or by a sponsor, which can be a commercial firm, a non-profit organization or the government (Porter 2004). The following table shows an additional differentiation criterion, the primary purposes of an OC, respectively of the participating members:

Primary purpose	Examples
Sell or buy products	Ebay.com, Hood.de
Social interaction	Facebook.com, XING.com
Advice seeking	Tripadvisor.com, KennstduEinen.de
Entertainment	Youtube.com, flickr.com
Knowledge and expertise exchange	Wikipedia.org
Problem solution	Gnome.com, mozilla.org

Table 1: Online communities classified according to their primary purpose

The range of purposes is broad starting from just trading products over communication with others to collaborative problem solving. Referring to the table above the expected innovativeness of the community increases from top to bottom. Nevertheless, a clear distinction of online innovation communities, representing the investigation objective of this work, from other OC follows in the chapter 2.2.2 Community-based innovation.

2.1.3 Modularity

More or less all entities, whether they are technological, biological, social or otherwise, can be viewed as a hierarchical system. This implicates that each entity, regardless of the analysis unit, turns out to be a system of elements and each of these elements is another system of elements until the lowest level of elementary subsystem is reached (Simon 1962). Elements are not mutually exclusive to only one system as hierarchies can overlap allowing elements to belong simultaneously to different systems. Abstractly, modularity can be seen as the degree to which the elements of a system can be coupled to each other, meaning divided and recombined. Considering that all systems have some kind of coupling between their single elements and completely undividable elements are extremely rare, more or less all systems are

10

modular, differing only in their degree of modularity (Schilling 2000). Modularity increases when a system emerges from a tightly coupled structure to a loosely coupled structure enabling a great variety of recombination between the individual system elements (Orton and Weick 1990). As pointed out previously, the concept of modularity can be applied for many different research disciplines, such as psychology, biology and even arts. For the context of this work, organizational modularity is most important and therefore explicitly described in the further proceedings. Nevertheless, organizational modularity is strongly influenced by modular technology (Hoetker 2006), as "organizations should be designed to reflect the nature of the tasks that they perform" (MacCormack et al. 2012, p. 2).

Considering the main strategic imperatives for organizations in the 21st century, modularity is one key organizational challenge. In recent decades, customized organizational designs were heavily promoted based on two beliefs (Nadler and Tushman 1999): First, as each organization faces unique demands, its design should be tailored accordingly. Second, the competencies gained within each designing process of the organization are valuable for the individuals involved and the entire organization in the long-term. Nowadays, due to reduced product life cycles, enhanced product complexity and intensified competition (Schilling and Steensma 2001), the organizational design needs to be radically changed towards a more modular design. From starting the design process with a blank sheet to an agreed "set of design principles that will allow organizations to quickly select an appropriated architecture for a given strategy"(Nadler and Tushman 1999, p. 55). Over the last two decades, many firms increased their level of modularity by replacing hierarchical entities by loosely coupled organizational components surrounded by semi-permeable boundaries (Schilling and Steensma 2001). As a result, the variety of contributors to a firm's production function is increased, the production locus is more often outside of the firm's responsibility, and management hierarchies within the firm are flattened. Specific characteristics of this transformation, respectively forms of modular organization are contract manufacturing, alternative work arrangements and alliances, for example. All three forms allow a loose coupling of required capabilities to the firm. Initiated by the electronics industry, many different industries rely rather on contract manufacturers to produce their designed products than on in-house manufacturing capacities (Plambeck and Taylor 2005). By using contract manufacturing, production capability can be loosely coupled to the firm, leading to a more modular organization and enabling faster responses to shifting demands and product changes. Alternative work agreements, such as temporary workers or contract agency workers, permit the firm to transform human capital into a loosely coupled component of the system. This organization form enables the firm not only to rapidly alter the workforce scale but also the mix of capabilities and talent required (Lepak and Snell 1999). Using strategic alliances can provide a firm with a wide range of capabilities that are loosely coupled to the organization. Especially financial and technological capabilities the firm is lacking can be accessed by

11

sharing risks of new ventures or a research and development agreement between two or more companies (Schilling and Steensma 2001). The previously detailed OC can also represent a loosely coupled element to the overall system. A diverse pool of volunteers provides a variety of capabilities for the firm without being tightly integrated into the organization.

Modularity features three major benefits: (1) Decreased communication costs, (2) economies of substitution, and (3) division of labor respectively specialization. Talking about the first benefit, communication consists of more than just pure information sharing or exchange. It can also represent the transmission of for example energy and materials (Baldwin 2008). The communication costs are significantly reduced in a modular system as shown in the next figure:

A non-decomposable system

	A	B	C	D	E	F
A	o	o	o	o	o	o
B	o	o	o	o	o	o
C	o	o	o	o	o	o
D	o	o	o	o	o	o
E	o	o	o	o	o	o
F	o	o	o	o	o	o

A nearly decomposable system

	A	B	C	D	E	F
A	o	o				
B	o	o				
C			o	o		
D			o	o		
E					o	o
F					o	o

A modular system with a common interface

	A	B	C	D	E	F
A	o	o	o	o	o	o
B	o	o				
C	o		o	o		
D	o		o	o		
E	o				o	o
F	o				o	o

Figure 2: Communication costs of different systems[8]

The first matrix represents a non-decomposable system in which each element (A-F) communicates with every other element leading to high communication intensity (Simon 1962). The (nearly) decomposable system showed in the second matrix has distinct lower community intensity but a full cooperation between all elements of the system is prevented. With the introduction of a common interface (element A) in the last matrix, a full cooperation among all system elements is guaranteed and the communication between all system elements is still reduced compared to the non-decomposable system (Langlois and Garzarelli 2008). Each element of the system only communicates with the elements in the same module and the common interface A, which acts as coordinator between the different modules.

For the other two benefits, economies of substitution and division of labor, the concept of economies of scope is of great importance. Economies of scope exist when the total costs of

---

[8] Following Langlois and Garzarelli 2008

producing two products combined are less than the sum of the costs when each product is produced separately (Helfat and Eisenhardt 2004). This cost reduction can occur for several reasons, e.g. a fixed production factor (most obviously a manufacturing plant but also other factors such as distribution channels) is not fully utilized by a single product, separate products naturally arising from the same input factor or the allocation of intangible assets among the products (Bailey and Friedlaender 1982). Teece (1980) identifies two main circumstances mostly responsible for the occurrence of economies of scope: "Two or more products depend upon the same proprietary knowhow base and recurrent exchange is called for, and when a specialized indivisible asset is a common input into the production of two or more products" (Teece 1980, p. 241). The economies of substitution concept is closely related to that of the economies of scope and "exists when the costs of designing a higher-performance system through the partial retention of existing components is lower than the cost of designing the system afresh" (Garud and Kumaraswamy 1995, p. 96). Similar to the previously mentioned reasons for cost reductions, this time the savings arise from the reuse of existing modules, the sharing of fixed investments and especially the knowledge associated with the retained modules. With an increased modularity level and standardized interfaces between the modules the economies of substitution rise as the retaining of existing system elements is facilitated. Probably the most important benefit from modularization, the favoring of labor division and specialization, is also mainly based on economies of scope. A high level of modularity simplifies the dividing of tasks, which can be assigned to the individuals or firms with the most competitive advantage in performing these tasks (Langlois and Garzarelli 2008). Due to its major relevance for this thesis, this modularization benefit is described in more detail in the following chapter.

2.1.4 Division of labor and specialization

In 1776, Adam Smith already recognized the importance of the division of labor for the wealth of nations gained from productivity growth: "The greatest improvements in the productive powers of labour, and the greater part of the skill, dexterity, and judgment, with which it is anywhere directed, or applied, seem to have been the effects of the division of labour" (Smith 1937, p. 5). He observed how a pin-maker divided the manufacturing process of this rather simple part into 18 different sub-tasks. By assigning each of these sub-tasks, sometimes grouped to a bundle of two or three sub-tasks, to one distinct worker, the daily output of the pin-maker increased significantly. Each of the men only needed the skills required to perform his distinct subtask(s), namely being a specialist for his job (Smith 1937). This specialization leads to comparative advantages respectively disadvantages between the parties involved in the manufacturing process. These differences resulting from specialization are the major rationale for international trade between countries in the past and nowadays (Harrigan 1997). Even though some workers have unequivocally intrinsic advantages in some tasks than other workers, Smith and others see the specialization of workers, meaning equipping them with the required skills and steadily improving these, as the main driver for gaining comparative advantages (Becker and Murphy 1992). With the steady development of new knowledge and ongoing technical changes, this investment in human capital is becoming more expensive due to the greater amount and complexity of the skills required. Vice versa this increases the need for specialization and is the main reason for a higher specialization degree in advanced economies (Rosen 1983). As Smith identifies the propensity of humans to trade as a reason for labor division, the market extends its dissemination: "As it is the power of exchanging that gives occasion to the division of labour, so the extent of this division must always be limited by the extent of that power, or, in other words, by the extent of the market" (Smith 1937, p. 10). In addition to Smith's argument for the curtailing of labor division, other reasons for limiting the degree of specialization have been identified. Especially the increase of coordination costs with an increasing specialization level is most frequently stated, as the group of specialized individuals to coordinate grows (Becker and Murphy 1992; Rodriguez-Clare 1996).

2.2 Innovation

"Innovation is composed of two parts, the generation of an idea or invention and the conversion of that invention into a business or other useful application" (Roberts 1988, p. 12). The source of innovation differs and is of paramount importance for innovation research and innovation management (von Hippel 1988). Since users are one key source of innovation, user and community-based innovation are outlined in this chapter. Afterwards empirical studies about member behavior in these innovating communities and the factors influencing their behavioral patterns are reviewed.

2.2.1 User Innovation

Since way back manufacturers, producers or more generally firms have been acknowledged as the major source of innovation. Starting in the 1970s this has been questioned more and more, when von Hippel first paid attention to the major importance of users in the innovation process by investigating innovation sources of scientific instruments (von Hippel 1976). "Users (...) are firms or individual consumers that expect to benefit from using a product or a service. In contrast manufacturers expect to benefit from selling a product or a service" (von Hippel 2006, p. 3). From those days, a significant research community has identified many cases in which intermediate[9] or consumer[10] users are the major driving force behind product, service or process innovations. Cases of intermediate users as a source of innovation embrace many different sectors such as machine tools (Lee 1996), plumber hardware (Herstatt et al. 1992), convenience stores (Ogawa 1998), library search and information systems (Morrison et al. 2000), and machinery automatic control systems (de Jong and von Hippel 2009). The majority of new product developments driven by consumer users take place in the field of leisure-time activities or sports-related goods (Bogers et al. 2010), such as mountain-biking (Lüthje et al. 2005), kite-surfing (Tietz et al. 2005) and sailing (Raasch et al. 2008), but also happen in other fields, e.g. even in sustainable energy technologies (Ornetzeder and Rohracher 2006). Even though most of the empirical studies identify consumer users as source for product innovation, service innovations can be driven by consumer users as well (Skiba and Herstatt 2009; Oliveira and von Hippel 2011). In some cases, the user innovator turns into a user entrepreneur by forming a firm and exploiting its development outcome. Contrary to the classic model of entrepreneurship, this process is rather emergent, meaning a step by step approach by starting to develop a product just for personal use without any

---

[9] "Users such as firms that use equipment and components from producers to produce goods and services" (Bogers et al. 2010, p. 859).
[10] "Typically individual end customers or a community of end users" (Bogers et al. 2010, p. 859).

purpose of selling it. Due to feedback loops with other users and further developments, a business opportunity is identified and the user innovator forms a firm (Shah and Tripsas 2007).

When trying to answer why users innovate, it all boils down to benefits and costs. Whoever expects to benefit maximally from innovating will be the source of innovation. Contrary to producers, users benefit alone from using innovations and not from selling the innovation (von Hippel 2006). Therefore, they more often develop fundamentally different solutions with high novelty tailored to their needs. These user innovators show characteristics of the so-called 'Lead Users'. These users, able to provide useful new product concepts to producers, also benefit significantly from innovations and face needs way ahead of others in the marketplace (von Hippel 1988). Complementary to the benefit aspect, the costs of innovation are often considered as an argument for the emergence of user innovation. While innovating several problems need to be solved and "to solve a problem, needed information and problem solving capacities must be brought together - physically or virtually - at a single locus" (von Hippel 1994, p. 429). Usually the need information is transferred from the user to the manufacturer who then in an iterative way develops products to satisfy the user's need. If this transfer is highly cost-intensive, the information about the user need is 'sticky', the innovation is more likely to happen on the user side (von Hippel 1994). When innovating, user innovators usually utilize information already in their possession, so-called 'local' information (Lüthje et al. 2005) further reducing the costs of innovating (Lakhani and von Hippel 2003). Baldwin et al. (2011) develop a model to explain if innovation might be rather driven by users or producers. For them communication and design costs are the major drivers for determining the locus of innovation. The higher the communication costs and the lower the design costs, the more likely a single-user innovator occurs. On the contrary, the higher the design costs and the lower the communication costs, the more likely the innovation model will be opened collaboratively driven by users. This form of innovation, often called community-based innovation, is the tenor of the following chapter.

2.2.2 Community-based innovation

Instead of innovating rather separately, new product developments increasingly take place in communities of volunteers (Franke and Shah 2003). These so-called Online Innovation or Collaboratively Innovation Communities, their major application field Open Source Innovation and the members of such communities are outlined in this chapter.

2.2.2.1 Online Innovation Communities

User innovation "rarely happens individually but rather requires interactions among like-minded" (Mahr and Lievens 2012, p. 169). User innovators are often widely distributed and therefore the sharing of problems, solutions or ideas among them is impaired. These information exchanges facilitate the development process and foster further innovations. Therefore, user innovation communities in particular or innovation communities in general also consisting of non-users fulfill an important role by making this information conveniently accessible. Innovation communities can be defined as "meaning nodes consisting of individuals or firms interconnected by information transfer links which may involve fact-to-face, electronic or other communication" (von Hippel 2006, p. 96). According to a previous definition, an Innovation Community is called OIC if the majority of the information transfer and communication takes place using the internet (Blanchard and Horan 1998). Additionally to an Online Community, the actors of OIC pursue "the clear purpose of contributing to the joint development of a single, integrated design or a number of interrelated designs. The design (...) is exploited in the sense that it is produced and sold on a market, integrated into other products that are marketed, deployed during the development of such products or used for any other private or commercial purpose" (Raasch et al. 2009, p. 383). OIC are a widely used instrument within Open Source Innovation (OSI) (von Hippel 2006), where the free revealing of information is an additional aspect.

2.2.2.2 Open Source Innovation

Incorrectly, OSI is often considered synonymously with Open Source Software (OSS), probably caused by the overwhelming amount of research work done exclusively for this stream. Some OSS projects are well known like Linux or Mozilla, but the phenomenon of OSI has already expanded beyond the software stream. The creation of content (Open Content) and even the development of tangible goods (Open Design) are drwaing more attention nowadays. Contrary to OSS, the stream Open Content covers all non-executable information goods, such as encyclopedias or multimedia (Cedergren 2003). Although most of the development work is done virtually, the purpose of Open Design projects is to design and finally produce a physical product (Raasch et al. 2009). This ranges from consumer products (e.g. beer) to hardware products (e.g. mobile phones and home entertainment systems) and even industrial goods (e.g. automobiles). Today's research provides various and overlapping definitions of OSI and OSS, but most recent definitions all embrace openness and collaboration. Collaboration has already been exposed when defining OIC, by stating the importance of the joint development of an integrated design by distributed actors. Regarding the innovation openness, Baldwin et al. state, "an innovation is 'open' in our terminology when all information related to the innovation is a public good - nonrivalrous and nonexcludable" (Baldwin and von Hippel 2011, p. 1400). In other words OSI contributors share their knowledge, designs, etc. openly among each other and the innovative outcome of this process is freely revealed for everybody to use. Thus, free revealing occurs when "all existing and potential intellectual property rights are voluntarily given up by that innovator and all interested parties are given access to it – the information becomes a public good" (Harhoff et al. 2003, p. 1754). Innovation openness is often considered as dichotomy, but it is rather a gradual concept (Balka et al. 2010) and detailed later in this thesis.

Regardless of the three streams, OSS, Open Content and Open Design, the research on OSI can be allocated to three different themes (von Krogh and von Hippel 2006):

- Motivations for contributing
- Governance, organization, and innovation process
- Competitive dynamics

The first wave of empirical studies of contribution motivation concentrates on identifying different individual motives for participating in OSI projects. Secondary research tries to investigate the impact for example firms' participation, community participation and technical design on the individual motives of the different actors involved. Almost contemporaneous research on the governance and organization of the OSI project as well as the innovation process itself arose. Especially the project architecture, the roles taken by individual project members and the coordination process are subjects of today's research. Research on the competitive dynamics resulting from the emergence of OSI has been slightly delayed when considering all three themes. The tremendous impact on the software industry and the relation between firms and OSI projects are investigated frequently. Additionally the resource allocation in detail and the involvement of commercial firms in general are of great relevance for OSI research (von Krogh and von Hippel 2006). A more comprehensive review of the literature on individual motives and community organization follows while proceeding. The members participating in OSI and OIC are manifold ranging from users to non-users and from individuals to corporations (Schweisfurth et al. 2011). Further characteristics of these community members will be explained in detail later.

2.2.2.3 Members of Online Innovation Communities

Prior research mainly characterizes members of OIC according to two dimensions, their role in the OIC and their demographics. An initial rough classification divides the community member roles into maintainers, developers/contributors and users (van Wendel de Joode et al. 2003). These groups can be further fragmented by their participation level or responsibilities (Xu et al. 2005). By analyzing four OSS communities in depth, Ye et al. identify the

18

presumably most granular range of community member roles (Ye et al. 2005):

Passive user: Passive users just using the innovative outcome of the community not attached to the community itself.

Reader: Readers actively participate in the community by reading mails, contributions, or source code.

Bug reporter: Bug reporters report errors or problems in the system but do not solve them. Therefore, they take some kind of testing role.

Bug Fixer: They fix the problem discovered by themselves or others and therefore need to understand and change a small portion of the system.

Peripheral Developer: These community members add new features to the existing system on a sporadic basis.

Active developer: They frequently contribute new functionalities to the system and are responsible for the majority of developments.

Core member: Core members also called maintainers, ensure the operational functionality of the community by coordinating and directing the development process.

Project leader: This person often initiated the community and drives the long-term development of the project.

Although this classification is based on OSS communities, a generalization to OIC from other fields seems to be appropriate. The four different community member groups of the project 'Wireless Leiden'[11] show a close similarity to the mentioned classification. The 'organizational user' initiating the community represents the project leader and 'volunteer users' installing infrastructure or maintaining the website stand for the developer or contributor role. 'Residential end users' subscribed to the wireless network and, sometimes giving feedback, act as passive users, readers or bug reporters. Finally the 'maintenance users' responsible for important nodes of the network show similarities to the core members (van Oost et al. 2008).

Diverse personal factors are frequently surveyed among members of OIC to determine their demographics and other characteristics. These factors embrace gender, age, education, professional background and nationality but also time spent on participating in OIC, previous

---

[11] "Wireless Leiden, a local wireless network infrastructure in the Dutch town of Leiden initiated, designed, and maintained by a local community of users" (van Oost et al. 2008, p. 182).

experiences with OIC or number of projects involved. The majority of the members are rather young to middle age (20-39 years) (Lakhani and Wolf 2005); (Ghosh et al. 2002) and primarily male (more than 95%) (Agerfalk and Fitzgerald; Hars and Ou 2001). Most of the participants live in the US and Western Europe (Lakhani and Wolf 2005) but Central and Southern American members seem to get more and more involved (Agerfalk and Fitzgerald 2008). The average community member holds a college degree or is currently enrolled at a university or college (Ghosh et al. 2002; Hars and Ou 2001). The majority has previous experiences in collaborative community development (Agerfalk and Fitzgerald 2008) and is participating in more than one project at the same time (Hars and Ou 2001). The community members can be divided into two main categories: 'commercial' and 'private'. Commercial actors, e.g., corporate users and manufacturers of complementary goods and services are often driven by the expectation of increased profit for their business (von Hippel 2007). Private actors often driven by the use value are usually not employed within the respective industry and financially not reimbursed (Hars and Ou 2001). Besides actors from these two categories, research actors can also be involved in OIC (Balka et al. 2009). A further noticeable fact about the surveyed community members is the relative high rate of self employees, freelancers or contract programmers (5-14%). These members are probably seeking business opportunities around the community's innovative outcome (Ghosh et al. 2002; Hars and Ou 2001).

2.2.3   Framing member behavior in Online Innovation Communities

As mentioned previously, the research on OIC is differentiated into motivation for contributing, community organization and competitive dynamics. Even though the different behavioral patterns of community members are no distinct stream, today's research does not completely lack this phenomenon. Primarily initiated in (von Krogh et al. 2003; von Hippel and von Krogh 2003; Lakhani and von Hippel 2003), the behavior of community members has recently attracted more attention. However, no explicit framework structuring the behavior of community members has been established so far and most of the publications regarding OIC and OSS still focus on the three defined research streams. Nevertheless, covered in the motivation and community organization stream, some publications primarily investigate the behavior of community members. To accumulate the research on this phenomenon a behavioral framework deduced from fourteen highly influential publications[12]

---

[12] Publications with focus on community member behavior identifying motivation to contribute and community organization only as explanatory element. In addition, these publications are published in high-ranked journals (A-level) and are intensively cited by others.

is derived in the next chapter. Afterwards the different identified behavioral categories and the factors driving member behavior are described in detail by reviewing additional literature.

The main research topic of this thesis, member specialization in OIC, is clearly part of the broader phenomenon of member behavior in OIC. The following chart shows the behavior framework derived to understand community member behavior comprehensively and the factors influencing this behavior:

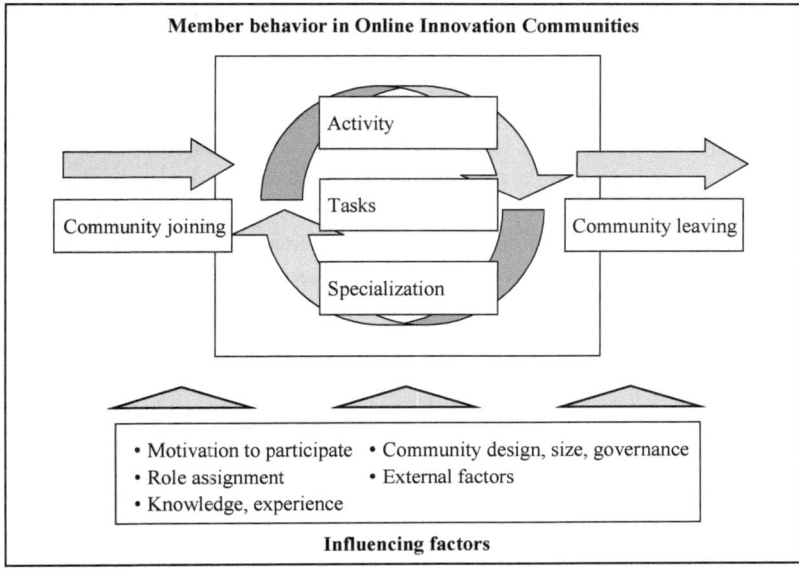

Figure 3: Framework member behavior in OIC

The derived framework clusters member behavior into five different categories accordingly the membership life cycle. Research on the first category *community joining* focuses on the circumstances under which the joining of new members is more likely (Butler 2001), the integration and development process of new members (e.g. von Krogh et al. 2003), and different joiner categories (Qureshi and Fang 2010). Publications addressing the second category *activity* mainly analyze how the average contribution activity differs among different member categories (e.g. Roberts et al. 2006) and especially the occurrence of lurking[13] and

---

[13] A Community member is defined as a lurker if he does not post or send a mail at all (Nonnecke and Preece 2000).

free-riding[14] (e.g. Baldwin and Clark 2006). Within the third category *tasks* different kinds of tasks performed by community members are identified (e.g. Lakhani and von Hippel 2003). The primary focus is to explain why certain members perform specific tasks and others do not (e.g. Shah 2006). Research is especially limited on the fourth category *specialization*. Today's research only distinguishes the existence of member specialization (von Krogh et al. 2003) but lacks the contribution focus of certain community members and the implications for the community caused by member specialization. Research on the last category *community leaving* mostly follows the first category and focuses on why and under which circumstances members decide to leave the community (e.g. Oh and Jeon 2007).

In addition to the five behavior categories, this framework also includes five influencing factors identified by research scholars while trying to explain the observed member behavior. Especially different individual motivations and characteristics of the community organization are mentioned most frequently. The knowledge or experience of the community members and if they occupy an assigned role in the OIC are additional factors driving the behavior. External factors such as the rise of new and similar Online Innovation Communities (Oh and Jeon 2007) are identified but do not take on any great importance when explaining member behavior. Other influencing factors such as group dynamics (Qureshi and Fang 2010), lead user characteristics (Jeppesen and Laursen 2009), and identification with the community and/ or company (Nambisan and Baron 2010) are rarely used. Before reviewing, the research on each behavioral category and the influencing factors in depth, the next table shows the fourteen highly influential publications and their contribution for developing this behavior framework:

---

[14] "The free-rider problem, also known as social loafing, occurs when one or more members of a group do not do their fair share of the work on a group project" (Brooks and Ammons 2003, p. 268).

Paper characteristics				Member behavior					Influencing factor				
Year	Journal	Authors	Title	Joi-ning	Acti-vity	Tasks	Specia-lization	Lea-ving	Moti-vation	Comm. orga.	Roles	Know., Exp.	Others
2001	Information Systems Research	Butler	Membership Size, Communication Activity, and Sustainability: A Resource-Based Model of Online Social Structures	x				x		x			
2003	Research Policy	van Krogh, Späth, Lahkani	Community, joining, and specialization in open source software innovation: a case study	x			x					x	
2003	Research Policy	Lakhani, von Hippel	How open source software works: free user to user assistance			x			x			x	
2003	Organization Science	von Hippel, van Krogh	Open Source Software and the Private collective innovation model: Issues for Organization Science		x	x			x	x	x		
2006	Management Science	Baldwin, Clark	The Architecture of Participation: Does Code Architecture Mitigate Free Riding in the Open Source Development Model?	x					x				
2006	Management Science	Shah	Motivation, Governance, and the Viability of Hybrid Forms in Open Source Software Development		x	x		x	x	x			
2006	Organization Science	Bechky	Gaffers, Gofers, and Grips: Role-Based Coordination in Temporary Organizations			x					x		
2006	Research Policy	Dahlander, Wallin	A man on the inside: Unlocking communities as complementary assets		x	x					x		
2006	Management Science	Roberts, Hann, Slaugther	Understanding the motivations, participation, and performance of OSS Developers: A longitudinal study of the apache projects	x					x			x	
2007	Management Science	Oh, Jeon	Membership Herding and Network Stability in the Open Source Community: The Ising Perspective		x			x		x			External factors, e.g. new OSS
2009	Research Policy	Jeppesen, Laursen	The role of lead users in knowledge sharing			x						x	Lead User character-istics
2010	Organization Science	Dahlander, O'Mahony	Progressing to the Center: Coordinating Project Work			x					x		
2010	Organizational Research Methods	Qureshi, Fang	Socialization in open source software projects: A growth mixture modelling approach	x					x			x	Socializa-tion group dynamics
2010	Organization Science	Nambisan, Baron	Different roles, different strokes: Organizing virtual customer environments to promote two types of customer contributions				x		x				Commit-ment, Iden-tification

Table 2: Publications in high-ranked journals with focus on member behavior in OIC[15]

In line with the chosen phenomena and concepts building the foundation of this thesis, the three dominating journals in this list are Research Policy, Management Science and Organization Science.

---

[15] Sorted according to year of issue

2.2.4   Member behavior categories

The following chapter outlines the five behavior categories, *community joining, activity, tasks, specialization,* and *community leaving* in detail. In addition to the fourteen publications used for identifying these categories other research contributions are incorporated in order to completely illustrate community member behavior.

2.2.4.1   Community joining

As this chapter comprises the behavior of community members in OIC, only the joining process once the member has already decided to join the community is described. It excludes the process of identifying, evaluating and finally choosing the community to join. The joining process of new community members and their different joining patterns are the major subjects when researching community joining. One of the most well known models concerning the process of joining an OIC is the onion model, in which members penetrate through the different layers into the core of the community (Crowston and Howison 2005). Transferred to an OSS community the layers are being a plain user, contributing to the mailing list, reporting and fixing bugs and developing at the core. Herraiz et al. (2006) point out that following this joining pattern towards becoming a core developer varies if the new member joins the community as a volunteer or as a hired developer: "Therefore, in short, we can say that the onion model is followed only by the volunteer developers in our sample, but not by those working for the project as employees. However, for all the developers following the model, the observed joining pattern is quite similar" (Herraiz et al. 2006, p. 32). Being promoted from a contributor to a code programmer, often defined as joining script, consists of different levels and types of activities. Observations of joining patterns in technical projects indicate that newcomers must show some level of technical expertise and at least a rough understanding of the project norms and rules before making any technical developments (Lovgren and Racer 2000). Scholars propose that contributors following this joining script are more likely to gain access to the developer community (von Krogh et al. 2003). The process of socialization with the core developers is an additional aspect of community joining and the lead-time for finalizing this process depends to a great extent on the initial level of social resources. Obtaining these resources can result from prior relationships to the core member of the community or the professional and educational background of the community joiner. A high initial level of these resources seems to accelerate the socialization but the process follows a non-linear increasing trajectory (Qureshi and Fang 2010).

Personal motivations, for example voluntary joined the community or assigned to do so, previous knowledge and experiences of joiners and different community characteristics play an important role for community joining. Butler observes a positive correlation between membership size and activity of the community with gaining new community members

24

(Butler 2001). A high proportion of experienced users in well established communities interferes with the joining process of new members: "These tendencies could be related to the continuing aggregation of experienced users as their activity and their proportion of the total activity increases more and more over time. It is difficult to reduce or at least delay these increasing imbalances, which make it hard for new inexperienced users to gain a foothold in the community" (Schoberth et al. 2003, p. 9).

2.2.4.2   Activity

Research about the members' contribution activity in OIC mainly tries to answer two different research questions: First, why are members inactive although they joined the community and do not contribute to the community? Second, which factors increase or decrease the average contribution activity of the participants?

The participation of Lurkers, community members not actively contributing to the community, only consists of consuming information by reading posts or mails (Curien et al. 2006). Preece and Nonnecke identify five main reasons for lurking (2004):

1. No need for posting, only browsing satisfies member's need
2. Time required to get to know the community before contributing
3. Being helpful by not posting so as not to distract the ongoing conversations
4. Technical problems, e.g. software installation unsuccessful
5. Disliking of community members, lacking the "belong to the group" feeling

Once a member has decided to participate actively, his contribution activity is influenced by a variety of factors. Specific personal motivations and the holding of a distinct role within the community can positively influence the member's activity. High affective[16] and continuance[17] commitment, individual characteristics deduced from their motivation show a positive influence on content provision activity. (Bateman et al. 2006) A comprehensive study by Roberts et al. indicates that being paid to contribute has a positive impact on the contribution intensity. This is additionally supported by Dahlander et al., who observe a high contribution activity of members sponsored by firms to spend a certain time contributing to the firm-hosted community (2006). Contrary, normative[18] commitment and surprisingly intrinsic[19] motivation

---

[16] Measured with e.g. " I feel like a part of the group (…)", "I have a strong sense of belonging to this site", "(…) real emotional attachment to this site" (Bateman et al. 2006).

[17] Measured with e.g. "(…) few alternative sites available", "The content of this site is too valuable for me to stop visiting", "(…) take me a long time to find a site that could replace it" (Bateman et al. 2006).

[18] Measured with e.g. " I would feel guilty if I stopped visiting the site now", "I felt an obligation to continue visiting this site", "I visit this site partly out of sense of duty" (Bateman et al. 2006).

[19] "Intrinsic motivation is defined as the doing of an activity for its inherent satisfactions rather than for some

seem to have no positive effect on the activity level. One approach to explain this observation could be that the intrinsic motivation of the participants is not fully aligned with the direction and mission of the community (Bateman et al. 2006; Roberts et al. 2006).

Factors affecting the activity negatively are strong external influences, a large group size and perceived unfairness within the social exchange process. In a simulation based on a theoretical model of Oh and Jeon (2007) an increase of external influences on the community members, such as the arising of new Open Source Communities or major changes in a participant's personal life, strongly decrease the average participation in small networks. In contrast, the participation decrease is less pronounced in big networks due to the high quantity of connections between community members within the network. Considering group size as a separate influencing factor, an increase in members goes along with an activity decrease by certain members: "When the group increases in size, the impact of any individual's participation in producing the collective good is negligible, and a self-interested, rational individual will choose to free-ride under these conditions" (von Hippel and von Krogh 2003). Increasing the degree of modularity seems to be one approach to prevent this behavior: "First, our model predicts that, ceteris paribus, open source codebases that are more modular or have more option value will attract more voluntary contributions (effort) than codebases that are monolithic or have low option value" (Baldwin and Clark 2006, p. 1126). A different study shows the importance of perceived fairness, affected e.g. by decision-making rights and ownership for the activity of community members: "The possibility of opportunistic ("unfair") actions by those holding control rights can both decrease and alter the character of volunteer participation" (Shah 2006, p. 1011). Especially the omission of the member's main contribution motive presumably always leads to a reduced activity since it has been observed that developers contributing due to use value reduce contributions once their problems are solved (Roberts et al. 2006). However, even alight changes in their motivations or learning goals as well as social status damage can also cause a reduced participation: "If users' motives change, the learning rewards have been exhausted, and/or the value of the social category depreciates, users are likely to reduce the level of their participation" (von Hippel and von Krogh 2003, p. 218).

---

separable consequence" (Ryan and Deci 2000, p. 56).

26

2.2.4.3 Tasks

Contributions to the community and several tasks performed by the community members can be substantially differentiated into information providing and information seeking. Information providers can be roughly[20] classified into 'pure contributors', expecting nothing in exchange for their contributions and driven by learning effects and reputation, and 'reciprocal contributors', motivated by the prospect of contribution by others. Information seekers can also be classified into two different categories: 'Pure askers', expecting information from other community members with nothing in exchange and the already mentioned 'Lurkers', consuming data randomly without asking for explicit information (Curien et al. 2006). A detailed study analyzing the tasks information providers perform shows a spending of 98% of their time screening and reading questions and answers from other community members and only 2% answering questions (Lakhani and von Hippel 2003). The surprisingly low effort in time for the provision results from exploiting knowledge already within the member's possession. The main benefit from the time-consuming screening and reading process seems to be learning achievements: "(...) information providers have had 98% of their effort rewarded via the learning they gain from scanning the questions and answers posted by others." (Lakhani and von Hippel 2003, p. 939). Jeppesen and Laursen (2009) identify the importance of low costs within the information providing process as well. Community members with lead user characteristics[21] tend to enjoy the revealing of their knowledge more than others. They are mainly driven by their low cost position as they use knowledge already available to them. Information seekers are also very important for the success of a community as a whole. They invest plenty of time to consume revealed knowledge suggesting to other community members the high value of their contributions. Nevertheless, they seem to show no special commitment to the community except continuance, which leads to the proposition that they are rather looking for private benefits than for larger social consideration of the community (Bateman et al. 2006).

In addition to the process of information providing and seeking, the kind of tasks necessary for the organization, coordination, success and viability of the community are frequently investigated. Therefore scholars identify many mundane but necessary tasks, such as provision assistance to users facing troubles with a product (Lakhani and von Hippel 2003), creating desired features and integrating contributions into the source code, respectively into

---

[20] Motivations to provide information to others in such an environment are manifold and explicitly explained in chapter 2.2.5.1 Motivation to engage in OIC.

[21] Measured with "I usually find out about new products and solutions earlier than others", "I have benefited significantly from early adoption (...)", "I have tested prototype versions (...)" (Jeppesen and Laursen 2009, p. 1585).

the product (Shah 2006), as well as triggering technical discussions (Dahlander and O'Mahony 2010). Observed coordination tasks embrace developing and propagating community norms (Bateman et al. 2006), the monitoring of community members to prevent free-riding (von Hippel and von Krogh 2003), and modularization and simplification of products and community organization (Shah 2006). Community members implicitly in charge of coordination and holding central positions within the network are often described as 'boundary spanners', 'organizational integrators' or 'brokers' (Burt 2005). They can link different sections of a community since they are well connected to different groups and are capable to ensure the successful knowledge transfer between them. These special individuals in the innovation process are characterized in more detail in the following chapter of this thesis.

Research scholars identify certain personal motivations why members perform these coordination and necessary tasks. Community members driven by enjoyment, learning and reputation rather than by their own need tend to stay longer with the community and perform coordination work (Shah 2006). Additionally they observe the phenomenon of shifting from a pure information provider to a boundary spanner with higher tenure (Nov et al. 2009). In their study the research scholars also confirm that extrinsic motivations like reputation and self development are positively correlated with performing 'mundane but necessary' tasks such as tagging photos (Nov et al. 2009). Tight personal ties with the community and its members are an additional argument for doing coordination work exceeding other participants (Bateman et al. 2006). In addition to different motivations, the assignment of certain roles to specific community members seems to enforce doing coordination work, respectively span boundaries. 'Becoming a project member' and 'being elected to a board director' significantly drive the participant to do technical and communication boundary spanning as well as participate in coordination discussions (Dahlander and O'Mahony 2010). Firm sponsored community members clearly dedicated to influence the community, so called 'Men-on-the-inside', grow into a central position of the network and hold strong communication bands to various members (Dahlander and Wallin 2006; Lee 2011). Bechky (2006) supports the importance of roles for the coordination process in OIC. He shows that in temporary organizations people have a clear role expectation of themselves and others and tend to fulfill these expectations implicitly.

2.2.4.4 Specialization

Only two identified major publications contain empirical studies about member specialization in OIC. These studies try to answer three different research questions: Do newcomers specialize on certain topics or modules? Do community members focus their contributions on a certain number of topics? When is a member more likely to contribute to the community and when to the community sponsor?

Von Krogh, Spaeth and Lakhani (2003) describe the distinction between specialization and generalization in OSS Communities as following: "High specialization indicates that the same modules within the code base were changed over time by a developer, while generalization indicates multiple modules were changed by a developer" (von Krogh et al. 2003, p. 1230). In their study, they observe a strong specialization of newcomers joining the community. Nearly sixty-six percent of the members entering the community chose only two out of fifteen modules for their first contribution. Furthermore, they support the evidence that participants focus their contributions precisely: "(...) we found evidence of high specialization. On average developers contributed to 4.6 (S.D. = 4.1) modules. However, 43% of all developers only contributed to up to two modules" (von Krogh et al. 2003, p. 1230). An initial approach to explain how participants focus their contributions assumes that community members specialize on the personally most rewarding topics at lowest costs for them (von Hippel and von Krogh 2003). The observation that core developers show a higher degree of generalization than the average community member, represented by contributing to many modules simultaneously, is explained by their long-term commitment to the project (von Krogh et al. 2003). Von Krogh et al. (2003) investigate only the specialization degree, meaning if community members allocate their contributions to more or less different topics. The contribution focus of members on specific content, respectively contribution types, is completely neglected. In a first empirical work focusing on different customer contribution types, Nambisan and Baron (2010) study the interdependency between different member characteristics or motivations and their main contribution type. They observe that a high sense of responsibility[22], the goal of self-image[23] and expertise enhancement[24] foster a contribution to the community, meaning sharing product expertise and answer peer customers' questions.

---

[22] Measured with e.g. "helping others (...)", "understanding peer customers (...)", "being a responsible and contributing community member (...)" (Nambisan and Baron 2010)

[23] Measured with „Enhance my status (...)", Reinforce my product related credibility (...)"(Nambisan and Baron 2010)

[24] Measured with e.g. „Enhance my knowledge about the product and its usage"," Enhance my knowledge about new product design features (...)"(Nambisan and Baron 2010)

In contrast to this, a high sense of partnership[25] and the goal of expertise enhancement foster a contribution to the sponsor, meaning collaborating with the firm in product development.

2.2.4.5 Community leaving

Compared to community joining community leaving is underrepresented in today's literature about community member behavior. Especially the detailed analysis of the leaving process is lacking. This is a surprising as the leaving of community members plays a major role within each project: "Most open and gated source developers (…) appear to leave the project within one year. Participants are not expected to remain on the project indefinitely and exit is understood to be a normal part of the process (…)" (Shah 2006, p. 1010). The fact that need-driven participants mostly leave when their needs have been met is an obvious motivation for leaving the community (Shah 2006). Other research scholars explain the leaving behavior with group dynamic effects. Oh and Jeon for example indicate the membership herding effect as a major cause for leaving the community. Each participant within the community network holds communication strings to a certain number of neighbors. When the majority of these neighbors decide to leave the community, the participant no longer experiences enjoyment or informal obligation and is likely to follow his direct neighbors (Oh and Jeon 2007). Butler also points out the importance of the group or community characteristics as he identifies an increase in member size, a higher contribution activity and a greater variation of topics as counterproductive to retain members (Butler 2001).

---

[25] Measured with e.g. „I understand how my contributions will be considered/utilized by the product vendor", "I generally receive quick reaction/feedback form the product vendor (…)"(Nambisan and Baron 2010)

2.2.5 Factors influencing community member behavior

As presented in the behavior framework, motivation for contributing, community characteristics, knowledge and role assignment are primarily used for explaining the different behavior of community members. For the successful prediction and explanation of the behavioral patterns unveiled in the empirical part of this thesis, a comprehensive understanding of these influencing factors is of utmost necessity and therefore they are detailed in the next chapter.

2.2.5.1 Motivation to engage in Online Innovation Communities

Voluntary contributing of knowledge, effort and time by individuals toward the collective benefit seems absurd, when they can free-ride on the others' efforts (Wasko and Faraj 2005). However, "why do top-notch programmers choose to write code that is released for free" (Lerner and Tirole 2001, p. 821)? Inspired by this essential research question from the OSI field, many researchers focus their efforts on identifying motivations why individuals contribute to OIC. As outlined in the previous chapter, these motivations are an important explanatory element for all five identified behavior categories of community members. Nevertheless, a conclusive discourse of all motivations and the according classification approaches identified by other researchers goes beyond the scope of this thesis. Therefore, eight main motivations constantly mentioned in literature are described briefly by listing some empirical findings:

Own Need

Participants face a need for e.g., a specific product feature or information and therefore decide to contribute to the community:

- "The need for software-related changes, alterations, or assistance drives initial and ongoing participation" (Shah 2006, p. 1005).
- The increased arising of serious ankle injuries inspired community members of a basketball community to jointly develop a better solution by mixing different shoe and insole models (Füller et al. 2007).
- Important drivers for participation are access to useful information, receiving of detailed help and acquiring of answers to complex problems (Wasko and Faraj 2000).

Enjoyment

Simply the fun and enjoyment when solving problems with like-minded people or developing new products, ideas or content drives members' participation:

- "Because I enjoy dealing with new products" (Füller 2006, p. 644) is observed as the main reason for initial and future participations.
- The reward motive "having fun programming" explains one major portion of overall participation (Hertel et al. 2003).
- Nov identifies fun as the main reason for writing or editing articles in the Wikipedia encyclopedia outscoring all other motivations (Nov 2007).

Enhancement

Enhancement embraces a wide range of different motivations. Two frequently mentioned are expertise enhancement, respectively learning, and career enhancement, meaning seeking career opportunities.

- Signaling their capabilities and skills by contributing to OSS projects to enhance their professional status is another motive to participate in OIC. Lahkani et al. (2005) observe this motive mainly for professionals and/or paid members.
- Further developing of skills, seeking opportunities to learn new things and enhancing expertise are important motives to participate in OIC. However, this reward motive can be seen as more important for joining OSS projects than Open Content projects (Oreg and Nov 2008).
- Investing in their own human capital by increasing their personal capabilities, knowledge and skills may be one reason for individuals' participation. Especially students and hobby programmers quote this motive as their main motivation (Hars and Ou 2001).

Reputation

Positive feedback and recognition by either peers or firms can positively influence the individual's choice to contribute to an OIC:

- "Innovative users may therefore feel proud when the firm acknowledges their innovative work openly in the community and perceive this recognition as an additional benefit of creating an innovation" (Jeppesen and Frederiksen 2006, p. 57).
- Research scholars support the hypothesis that reputation gain motivates individuals to contribute their knowledge to a network of others (Wasko, Faraj 2005).
- Community participants motivated by reputation building tend to invest plenty of time to strengthen social relationships to their community (Nov et al. 2009).

Reciprocity

Reciprocal behavior can be defined as "Contribute valuable information to the group in the expectation that one will receive useful help and information in return" (Kollock 1999, p. 227). In contrast to direct reciprocity, meaning receiving help from the same individual in exchange for my help, generalized reciprocity is more likely to occur in an OIC setting. Generalized reciprocity means that individuals expect to receive help from anyone out of the group or from the group as a whole and not from a specific individual (Wasko, Faraj 2000):

- The importance of the reciprocity motive seems to decrease when providing help more frequently to other community members (Lakhani and von Hippel 2003).
- The engagements of individuals based rather on their affiliation towards the community than on private benefits are often explained by reciprocal factors (Franke and Shah 2003).

Financial compensation

Community members motivated by financial reimbursements for the time and effort they spend on contributing to an OIC can be separated into two categories. The first group consists of participants, professionals or hobbyists, who are paid to join the community. The second group contributes to the community and hopes to receive some financial compensation in return for their work. For the second group motive researchers are still struggling to find evidence that financial compensation motivates participation (Lüthje 2004; Lakhani and Wolf 2005; Shah 2006). Nevertheless, some empirical findings show at least a weak positive effect of monetary rewards, e.g. in terms of participation frequency (Füller 2006).

Efficacy

The act of contributing valuable information to an OIC can result in a sense of self-efficacy, meaning that the individual senses he or she has a positive effect on the environment. "Making (...) contributions to the group can help individuals believe that they have an impact on the group and support their own self-image as an efficacious person" (Kollock 1999, p. 228):

- Participating in projects seems to give community members a strong feeling of effectiveness, competence and accomplishment (Hars, Ou 2001).
- Writing and editing articles in the Wikipedia online encyclopedia seem to give individuals a sense of being needed and a high motivation level. Self-efficacy correlates significantly with a higher contribution (Nov 2007).

Altruism

The definition of altruism changes slightly depending on the specific discipline such as sociobiology, economics or sociology. On a higher level, it can be defined as "acting with the goal of benefitting another" (Piliavin and Charng 1990, p. 27) or "doing something for another at some cost to oneself" (Ozinga 1999, p. 5). Altruism is extensively used for explaining the occurrence of voluntary giving and donations of any kind, such as time, blood and money (Smith et al. 1995; Houston 2004). Altruism is also important for profiling green consumers, meaning characterizing consumers who are inclined to buy ecological products[26]. Many researchers observe that altruistic persons show a higher willingness to buy these kinds of products (Roberts 1996; Rowlands et al. 2003; Menges et al. 2005). Donating time and effort to benefit others at some cost to oneself, namely contributing to an OIC, indicates altruistic behavior and therefore altruism can be one motive to participate. Participants in various surveys among OIC members frequently state altruism as one motive for their engagement (Wasko and Faraj 2000). Altruism seems to be more important for participating in Open Content communitites rather than in OSS communitites (Oreg and Nov 2008).

Commonly motivation is classified into intrinsic and extrinsic. Contrary to intrinsic[27], "extrinsic motivation is a construct that pertains whenever an activity is done in order to attain some separable outcome" (Ryan and Deci. 2000, p. 60). The allocation of the described eight motivations to these two main classes differs among different researchers and disciplines. As a mutually exclusive allocation is highly questionable and a clear determination is not necessary for the proceeding, a differentiation is not executed at this point.

2.2.5.2 Community characteristics

As seen in the premises of this thesis, the characteristics community organization and members are often utilized to explain user behavior in OIC. Frameworks clustering OIC according to different characteristics including e.g. initiator (firm sponsored or community founded) and interactivity (none, indirect or direct) are well established in today's research (Tietz 2007) and have already been discussed briefly earlier in this thesis. Due to the focus of this study, only the community characteristics most important for participant behavior are explained in detail based on the following classification:

---

[26] e.g.,electricity from renewable sources, cars with reduced emissions, locally produced products, compensation of CO_2 emissions, etc.
[27] Already defined in the chapter 2.2.3 Member behavior in OIC

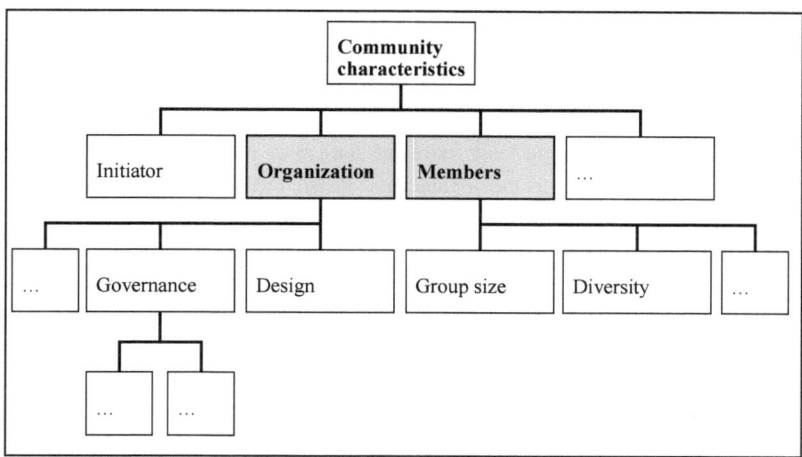

Figure 4: Community characteristics influencing member behavior

Community governance

Markus identifies the rules and structures to manage the community and the control of the innovation process as two main categories of community governance studied in prior research (Markus 2007). Community management contains e.g. the installation of central authorities as control mechanisms to increase success and efficiency (Galivan 2001), as well as the guidance to manage the community boundaries, meaning rules for obtaining membership status and identification (O'Mahony and Ferraro 2004). Studies about the innovation process in OIC mainly focus on the differences to the existing innovation theory derived from firm-based product development. The product development approach in communities often follows incremental changes and improvements but does not support complex new feature development (Jorgensen 2001). Taking this difference into account, the development process in OIC needs to be managed differently: "(...) the product development process can be effectively organized as an evolutionary process of learning driven by criticism and error correction" (Lee and Cole 2003, p. 633).

Especially in Open Source Communities intellectual property and the degree of openness are important aspects of community governance. Although project contributors freely reveal their work and make it publicly available, they do not give up their rights to it (O' Mahony 2003). First, the fundamental question of ownership needs to be answered: Who owns the content of the project and derivative subprojects, the sponsor of the community or the community and its members themselves (West and O' Mahony 2008)? Second, choosing the right license type is essential for managing the community content successfully. In prior research, Sen et al.

categorize the different Open Source license types into Strong-, Weak and Non-Copyleft. The Strong-Copyleft License is the most restrictive form, where derivative software based on the original must be licensed accordingly to the original. The Weak-Copyleft can be considered as moderately restrictive and allows the linkage with other non OSS licensed programs. The Non-Copyleft license and therefore least restrictive form does not regulate the license of derivative software at all as long as the original developers are credited for their underlying work (Sen et al. 2008). Projects mainly directed towards end users, e.g. games, tend to be more restricted whereas projects geared to developers and running on commercially operating systems show a less restrictive license (Lerner and Tirole 2005). The openness of an Open Source project consists of three different aspects. Balka et al. extend the framework proposed by West and O' Mahony (2008) sketching the aspects transparency and accessibility by the additional aspect of replicability for analyzing the OSI phenomenon beyond software. "Transparency refers to the quantity and quality of information which is freely revealed to developers. (...) Accessibility denotes the possibility for community members to actively participate in product development. (...) Replicability denotes the availability of individual components and thus the possibility for the self-assembly of the product (...)" (Balka et al. 2010, p. 4).

Community design

As pointed out previously, research scholars mention the degree of modularity of an OIC as one aspect to explain certain behavioral patterns of the community members. Modularity as an element of community design helps to answer the question how OIC successfully coordinate the collaboration and knowledge sharing process (Langlois, and Garzarelli 2008). Narduzzo and Rossi (2005) show that both the organization of the project (community) itself and the innovation objective (product) benefit from the modularity paradigm. The general benefits from modularity, limitation of redundant communication and simplified coordination, the economies of substitutions, and the enabling of labor division and specialization all affect OIC. The communication intensity and therefore the communication costs in a non-decomposable system compared to a modular system with common interfaces are distinctly higher (Langlois and Garzarelli 2008). Utilizing the economies of substitutions, developers of OSS projects for example reuse code already written for other projects to lower development costs, gain additional time for more favorable tasks and introduce new functionalities and features more quickly (Haefliger et al. 2008). Lastly, the division of labor facilitates the application of often extremely specialized knowledge which is already available, as user innovators in OIC tend to do (Jeppesen and Laursen 2009; Lakhani and von Hippel 2003).

Membership group size

An increase in membership group size seems to be favorable for all kinds of communities. The resources of an OIC are its members, more precisely their knowledge and effort, therefore the size of a structure's membership group indicates the amount of available resources. In large social systems it is more likely that one member knows a needed solution right away, has time to coordinate the collective effort or can provide any social support (Butler 2001). Additionally individual participants in OIC looking for e.g. reputation from peers and career enhancement are likely to prefer a larger audience to a smaller one. Consequently, membership group size has a positive effect on attracting new community members.

Additionally a larger group size helps to overcome start-up problems, since a critical mass is needed to sustain a long term interactive discourse in interactive media and therefore OIC (Markus 1987). However, contrary to prior critical mass model research, Butler (2001) shows that an increase in membership size also has negative effects. An increased number of users cause two major challenges. First, "large systems weaken the possible linkage between contributor and beneficiary, diluting the hope of reciprocity" (Thorn and Connolly, p. 521). Vice versa this leads to fewer participant contributions and more social loafing, lurking and free riding. Second, an increased membership size leads to an increased communication volume and can finally result in a state that is called 'information overload' (Schoberth et al. 2003). Community members therefore reduce their activity or leave the community entirely. To prevent this, certain upper limits for the number of participants actively involved in the community are required based on "(...) the technology type and the cognitive effort that participants are prepared to devote to message processing" (Jones and Rafaeli 1999, p. 246).

Membership diversity

In communities two different types of heterogeneity can occur: interest and resource heterogeneity. Even in groups with similar interests, as usually assumed when considering OIC, interest conflicts appear. Some members might have a different understanding of the basic group interests or their interest level is greater or smaller on certain topics compared to the other community members. Resource heterogeneity appears when the resources available to the community members differ substantially (Oliver et al. 1985). Similar to membership group size an increase in group heterogeneity seems to be favorable. A high level of different resources available to the group members and therefore to the community as a whole increase the opportunity to e.g. solve problems more effectively and faster. Early studies deliver evidence for the favorability as "(...) group answers of superior quality (were observed) when the group was composed of heterogeneous individuals rather than homogeneous individuals" (Aamodt and Kimbrough 1982, p. 171).

However, a high level of heterogeneity is dichotomous since it can also split the community into opposing groups. Later more explicit research that is more explicit indicates that heterogeneity fosters collective action at an initial status when problems tend to be unstructured but that homogeneity promotes the maintaining of established systems when problems are well structured (Heckathorn 1993). A high level of heterogeneity can lead to asymmetries within the group and limits knowledge sharing, respectively information exchange. Unequal information benefit, namely interest heterogeneity, and unequal information quality, namely resource heterogeneity, jeopardize the exchange between group members (Thorn and Connolly 1987). A low information benefit compared to other group members seems to be one reason for lurking, since lurkers frequently cite that the joined group has no or limited value for them (Preece et al. 2004).

2.2.5.3 Knowledge (sharing)

The three different perspectives on knowledge are "knowledge as the object, knowledge embedded in individuals, and knowledge embedded in community" (Wasko and Faraj 2000, p. 157). When considering knowledge as the object researchers conceive it as some graspable truth, which is not dependent on someone's mind and can be codified (Townley 1993). It is easily transferable to organizations and individuals regardless of the person that contributed the knowledge in the first place. Contrary to the next perspective, the knowledge as the object view assumes that an individual does not own knowledge.

Considering the arguments of other researchers in the previous chapter about community member behavior in OIC, the second perspective knowledge of individuals is a paramount influencing factor. User innovators often rely on 'local' information that is already in their possession (Lüthje et al. 2005). Community members also utilize knowledge that is already available for their contributions (Lakhani and von Hippel 2003; Jeppesen and Laursen 2009). This knowledge embedded in an individual's mind also defined as expertise can be described "(...) as a human quality that builds on data and information together with experience, values, and insight" (Rafaeli and Raban 2005, p. 63). Individual knowledge can be differentiated into tacit and explicit knowledge[28]. Explicit knowledge can be articulated into text, word and signs whereas tacit knowledge is rooted in the owner's head and therefore often not easily transferable (Sowe 2008). Interaction with other individuals and thereby externalization of one's own tacit knowledge is one possibility to at least partly transform tacit into explicit knowledge (Nonaka 1994). One risk for organizations like OIC is that this individual knowledge is not easily convertible into a structural asset, meaning that this knowledge can

---

[28] Tacit-explicit is only one dimension of knowledge, others e.g. universal-local, declarative-procedural exist but are not considered here as they are of minor importance for this study.

be completely lost if the individual decides to leave the organization (Wasko and Faraj 2000). However, not only the participants of OIC can hold valuable knowledge but also the community itself, which is explained in the following perspective on knowledge.

The third perspective, knowledge embedded in communities, assumes that knowledge is maintained by the community and owned collectively (Wasko and Faraj 2000). Knowledge creation is jointly accomplished through open discussions and collaborations between different actors using e.g., mailing lists or chats. This perspective covers a wide range of knowledge forms, such as common language, routines, meanings and the commonality of specialized knowledge (Grant 1996). Community knowledge gaining, respectively learning, happens not merely from observation but also from developing a common understanding through interaction (Brown and Duguid 1991). Especially the conversion of tacit into explicit knowledge is simplified when a common understanding between the individuals involved has been established. This knowledge integration into the organization is particularly important if community members come from non-identical expertise fields but obsolete if they possess identical knowledge (Grant 1996).

As previously outlined in this thesis for a high number of community members, participating in OIC means to share their knowledge. Knowledge sharing implies a relationship between at least two parties, one who possesses the knowledge and the other one who acquires the knowledge. "The first party should communicate its knowledge, consciously and willingly or not, in some form (…). The other party should be able to perceive these expressions of knowledge, and make sense of them (…)" (Hendriks 1999, p. 92). The knowledge sharing process within an OIC typically involves a request for help or an open question by the knowledge seeker and in response the sharing of knowledge by the knowledge provider (Hara and Hew 2007). Researchers often quote the terms 'externalization' and 'internalization' when analyzing knowledge sharing activities in OIC. 'Externalization' describes the process of transferring expertise or knowledge to a knowledge management system, where the knowledge nugget is stored. 'Internalization' is the process of acquiring knowledge from this system (Sowe et al. 2006). These knowledge management systems are information systems designed to enhance knowledge sharing by lowering barriers between knowledge seeker and provider such as space, time, social distance or language (Hendriks 1999). Mailing lists, discussion forums, chat rooms or whiteboards are typical information and communication technologies serving as a knowledge management system supporting the knowledge sharing and exchange process in OIC (Wasko and Faraj 2000).

2.2.5.4 Role assignment

Considering the characteristic of roles in the environment of an OIC, Baker et al. provide a pleasant role definition. Referring to the structural role theory they define a role "as a bundle of norms and expectations - the behavior expected from and anticipated by one who occupies a position (or status) in a social structure" (Baker, Faulkner 1991, pp. 280–281). Sometimes position and role are used synonymously, which can cause some serious confusion. Whereas position stands for a location of an individual within a certain social structure, role is rather a classification of individuals across different social structures (Winship and Mandel 1983). Enactment connects roles and position and in the traditional view the role of an individual is enacted from the position the individual holds within a social structure (Turner 1985). Nevertheless, recent research on temporary organizations and OIC unveiled the reverse direction of this enactment. Roles are assigned to specific individuals, which afterwards grow into distinct positions (Bechky 2006; Dahlander and Wallin 2006; Lee 2011). Turner (2003) provides a possible explanation for this behavior: "By virtue of playing a role, individuals incorporate meanings and expectations associated with this role into their identity in the situation" (Turner 2003, p. 376).

2.3 Research questions

Considering the literature review about community member behavior in OIC previous research primarily lacks the behavior category of member specialization. Von Krogh et al. (2003) investigate if newcomers specialize on certain modules and by additionally determining the degree of specialization show initial findings that core developers generalize and other community members specialize. Nevertheless, the specialization degree already mentioned by Smith in 1776 (Smith 1937) is only one aspect of specialization. Another one is what kind of contribution is likely to be made by which community member. In other words, what kind of content is most likely contributed by certain community members? The only other known study investigating specialization behavior picks up this aspect by trying to answer if community members tend to contribute rather to the community or the company (Nambisan and Baron 2010). Another recently published work by Mahr et al. (2012) investigates the focus of lead users' contribution in virtual lead user communities. However, they show what type of contribution is more valuable to the community and not the contribution focus of the users themselves. Nambisan and Baron (2010) emphasize the lack of research on specialization, especially on the contribution focus: "To our knowledge, this is the first empirical work that has explicitly focused on different types of customer contributions in online customer forums (…)" (Nambisan and Baron 2010, pp. 566–567). Other researchers also identify a need for further studies in this direction: "Research is still lacking on the benefits of specialization in open source software innovation (…)." (von Krogh et al. 2003, p. 1218) or "future studies should also investigate whether or not developers change their degree of specialization over time (…)" (von Krogh et al. 2003, p. 1235). Even though Nambisan and Baron manage to identify explanations for the member affinity towards certain contributions, more research seems to be required.

Research question 1: How do community members focus their contributions?

Considering the research of member behavior in OIC, the majority of the published studies cover more than just one of the behavioral categories. The simple reason is that the behavioral categories influence each other, e.g. a community member intensively contributing performs different tasks and also displays a lower degree of specialization with a simultaneous high activity level (von Krogh et al. 2003; Shah 2006). Nevertheless, the impact of specialization on other previously mentioned behavioral categories, such as activity and community leaving, is unexplored. For example if the content of the community discussion alters, the likelihood that community members focused on a certain content act differently from other community members is rather high.

Research question 2: How does an altering community focus influence the contribution activity of specialized members?

Besides the lack of research on the community members' contribution focus and its influence on their activity level, the implications of specialization for managing OIC successfully are completely lacking. OIC riddled with specialized members face two major challenges. First, specialized community members hold individual tacit knowledge, which needs to be transferred into explicit knowledge to benefit the community (Nonaka 1994). Second, as outlined in the chapter 'Knowledge (sharing)' knowledge attached to individuals is completely lost when the individual decides to leave the community (Wasko and Faraj 2000).

Research question 3: How is knowledge from specialized individuals integrated permanently into the community?

3 Theoretical background

This chapter outlines general theories applied to predict and explain the behavior of community members and closes with deriving the research hypotheses employing the previously drawn theories.

3.1 Theories of individual behavior

In this chapter important elements of two major theories of individual behavior, the rational choice and the social cognitive theory, are outlined. Both theories are carefully selected. The rational choice theory postulates the maximization of utility that again is the major driving factor beyond labor division and specialization. As outlined previously, the behavior of community members is influenced by personal and environmental factors. This again is postulated by the social cognitive theory. Finally, key individual roles in the innovation process, e.g. boundary spanners and promotors, are described in detail and their importance for OIC is discussed briefly.

3.1.1 Rational choice theory

The rational choice theory (RCT), also called choice theory of economics or optimal choice theory, is based on the utilitarian economics that were initially formulated by Adam Smith in the 18th century. According to him, "actors are conceptualized as rational and as seeking to maximize their utilities" (Turner et al. 1991, p. 325). From a purely economical view this reflects the maximizing of benefits by simultaneously minimizing costs. The underlying idea of RCT is that individual behavior consists of numerous decisions made by individuals trying to maximize their utility. In microeconomics, the concept of utility maximization is often illustrated by the optimal consumer choice, where an individual choices the basket of goods maximizing his or her utility (satisfaction) (Besanko and Braeutigam 2007). The RCT makes two basic assumptions about the decision process of an individual. The first one is completeness, meaning that all possible actions can be ranked according to their values (two or more actions with the same value are possible). The second assumption transitivity secures the consistency of the actions' rank. This implies that if action A has a higher value than action B, and action B has a higher value than action C, action A is also preferred to action C (Tversky and Kahneman 1986). Although the RCT is doubtlessly originated from economics, "all the disciplines dealing with behavior, from political philosophy to behavioral biology, rely increasingly on the idea that humans and other organisms tend to maximize utility (...)"

(Herrnstein 1990, p. 356).

In studies of individual behavior, applying the RCT is of unique attractiveness due to its usefulness for explaining a great variety of individual actions. However, critics doubt the explanatory contribution of RCT as the rational action itself serves as explanation (Hollis 1977). Even though many research scholars frequently quote the utility maximizing concept[29] for explaining behavior of OIC members, the RCT should only be applied as a basis for investigating individual behavior. The rational aspect behind individual's choices is more complex and factors exceeding the rational utility maximization need to be considered. "Rational choice theory tells us how choice should be allocated, given a reinforcement or utility structure, not how it will be allocated" (Herrnstein 1990, p. 366).

3.1.2 Social cognitive theory

Three major approaches in behavioral research, naturally including individual behavior, can be identified (Davis and Luthans 1980):

1. Behavior $= f$ (person). The behavior of an individual is predicted and explained by using personal characteristics such as perceptions, attitudes and motivations.
2. Behavior $= f$ (environment). According to this second approach, differences in individual behavior are explained by using external variables (Pfeffer et al. 1976)
3. Behavior $= f$ (person, environment). The third approach compromises that both the person and the environment determine the behavior of individuals. Especially the interaction between these two factors significantly drives the behavior and neither of both should be overemphasized when developing behavioral models (Chatman 1989).

The social cognitive theory postulates the latter approach by assuming a recursive interaction between individual behavior, personal factors and the environment (Bandura 1988). In the social cognitive theory or the related social learning theory, "(...) behavior is viewed as affecting and being affected by the participant's cognitions, the environment, and the person-situation interactions" (Davis and Luthans 1980, p. 283). The social cognitive theory provides a framework for understanding, predicting, and changing human behavior (Bandura 1977). The following figure adapted from Davis and Luthans (1980) graphically shows the triangle model of the social cognitive theory:

---

[29] E.g. von Hippel and von Krogh (2003) observe that community members contribute accordingly to maximize their benefits and minimize their costs.

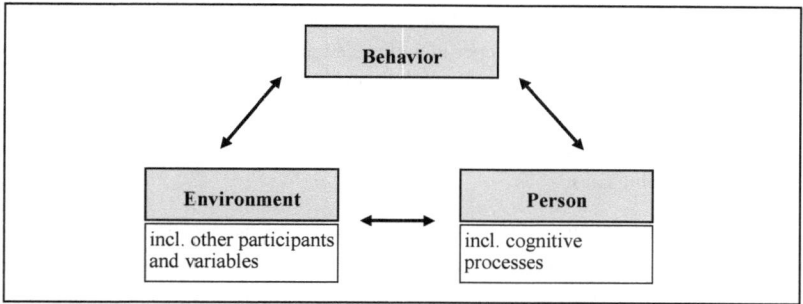

Figure 5: Model of social cognitive theory of organizational behavior[30]

The previously identified factors influencing the behavior of OIC members can be classified into personal and environmental. The individual motivation of the member for engaging in the community, the member's knowledge or previous experience and the occupation of a role are related to the member or person. On the contrary the community design and composition as well as external factors are undoubtedly environmental factors driving the individual behavior of community members.

3.1.3   Theories of key individuals in the innovation process

Organizations are encircled by internal (within the organization, e.g. between sub-units) and external boundaries (between different organizations). In order to innovate constantly firms need to generate, acquire and integrate external and internal knowledge into their organization (Rosenkopf and Nerkar 2001). Therefore, firms need to overcome the communication barriers between these different boundaries (Tushman and Scanlan 1981). Hence, certain individuals in the innovation process are of paramount importance for the diffusion of external knowledge into and internal knowledge through the organization, namely technological gatekeepers[31] or the related concept of boundary spanners. In the majority of cases, technological gatekeepers are firm members located within the R&D department. These key individuals "differ from their colleagues in their orientation towards outside information sources" (Allen 1970, p. 15). They are more likely to read scientific journals or professional engineering magazines than others and frequently hold strong communication ties to external technologists. One key function of the technological gatekeeper is to mediate between the

---

[30] Referring to Davis and Luthans 1980
[31] Other critical roles of individuals in the innovation process are e.g. idea generator, program manager or sponsor (Roberts and Fusfeld 1981)

world outside of his or her own organization and their organizational colleagues (Allen 1970). In a first step, the technological gatekeeper identifies and analyses information from outside the organization. This is followed by communicating the information into the organization in such a way that other organizational members can easily process this information (Tushman 1977). Hence, with an increased specificity of the information required, the importance of technological gatekeeping rises.

Although the exact differentiation between gatekeepers and boundary spanners varies among different scholars, a clear relation between these two concepts cannot be denied. These individuals located at the boundaries of an organization "exist to mediate communication across several organizational interfaces" (Tushman 1977, p. 602). As mentioned previously, these interfaces can exist within an organization or between organizations and their environment, therefore boundary spanning can be classified into external and internal (Rosenkopf and Nerkar 2001). Contrary to technological gatekeepers, boundary spanners are not limited to the transfer of technical information, but can also gather and transfer all kinds of information (Reid and de Brentani 2004). Similar to technological gatekeeping, this transfer is a two-step approach. Initially they screen, analyze, filter and decide what information is of major importance for the organization. Therefore, boundary spanners should feature a high expertise level in the required field. According to the concept of absorptive capacity the ability to fulfill this task is strongly positively influenced by the knowledge level of the individual in the respective field (Cohen and Levinthal 1990). The second step is the distribution of the gathered and filtered information into the organization. The success of this "dual function in information transmittal" (Aldrich and Herker 1977, p. 218) strongly depends on the expertise of the boundary spanner to decide which unit in the organization requires this information and in which manner. In addition to their information processing function, boundary spanners can also fulfill other important tasks, such as representing the firm externally (Aldrich and Herker 1977). Research on boundary spanning initially started by observing individuals within R&D departments of firms (Tushman 1977; Tushman and Scanlan 1981). However, the literature has been expanded to incorporate boundary spanning activities between firms (Rosenkopf and Nerkar 2001) and within online communities (Fleming and Waguespack 2007). Especially in big and highly modularized online communities, boundary-spanning activities are essential. Their organizational boundaries are manifold (between different community modules, to other communities, to the sponsor, to the environment), the number of knowledge carriers is rather high and these are widely distributed among different locations.

In addition to the previously mentioned organizational barriers, several others such as psychological and capability barriers jeopardize the success of the innovation process (Witte 1999). The promotors-model claims that a "cooperation of several different kinds of specialized promotors" (Rost et al. 2007, p. 341) rather than one individual can jointly cross these barriers to guarantee successful innovating. Considering this definition, the model represents a more comprehensive concept of key individuals in the innovation process than the concept of gatekeeping and boundary spanning. Promotors "actively and intensively support innovation processes" in organizations (Witte 1977, p. 53) and scholars identify four different types of these key individuals:

1. Power promotors
2. Expert promotors
3. Process promotors
4. Relationship promotors

Power promotors actively drive and support the innovation process through their hierarchical potential (Hauschildt and Kirchmann 1999). This potential is partly based on their position within the organization and partly to their specific behavior. Their position needs to empower them to sanction preventers or opponents of innovation and to reward the individuals showing innovation willingness. Power promotors are not characterized by simply tolerating the innovation process with some kind of passive goodwill (Witte 1999). Rather, power promotors actively foster innovation throughout the entire organization and hold strong communication ties to the organizational members in charge of developing innovations. Therefore, resistance towards innovation is likely to occur when the knowledge of the individuals engaged in or affected by this innovation is simply insufficient. To overcome this innovation barrier of 'not-knowing', expert promotors equipped with specific knowledge need to act as teachers or technologists to make other organizational members adopt the innovation (Hauschildt and Kirchmann 1999). The hierarchical position of expert promotors is completely irrelevant and they can gain their expert knowledge either from work or from their personal affinity towards the novelty. Expert promotors do not only fulfill the function as knowledge carriers but also as convincers of others to finally use the innovation (Witte 1999). Process promotors are very familiar with the intra-organizational structures and know which individuals within the organization are affected by the innovation. They perform a linking function between power and expert promotors and are capable of facilitating novelties to other organizational members explicitly in the required form (Hauschildt and Chakrabarti 1999). Therefore, the role of the power promotor is similar to the previously outlined internal boundary spanner or gatekeeper role (Hauschildt and Schewe 1999). The role and tasks of relationship promotors show close similarities to those of the process promotors. Nevertheless, the innovation process benefits from their inter-organizational knowledge that is represented by their network to suppliers, customers and individuals from other innovating organizations (Gemünden and Walter 1999). This kind of external boundary spanner unites the right process participants at the right time to overcome barriers hindering innovation.

Following Gemünden (1999), the efficiency of the innovation process can be significantly increased if promotors are heavily involved. Especially the contribution by more than one promotor type, whether paired up either as team or the same person fulfilling multiple roles[32], is favorable for the innovation outcome. The following figure shows some contribution types made by expert, process and power promotors supporting innovative activities within the organization:

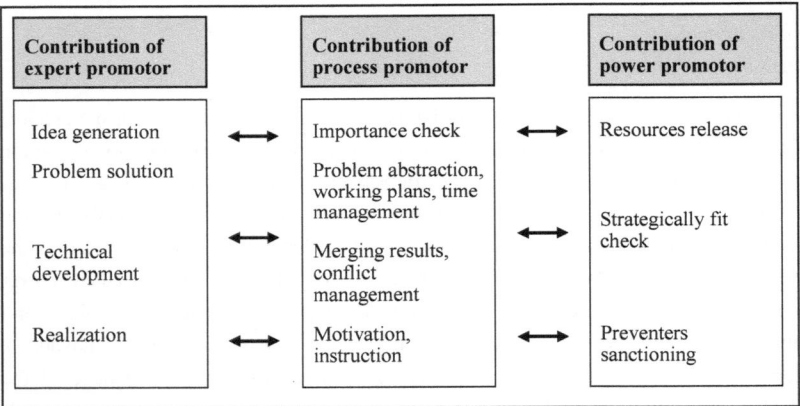

Figure 6: Contributions of promotor roles along the innovation process[33]

Considering OIC, where innovation-affine volunteers commonly develop new products or services and often freely reveal their ideas and knowledge, the opponents or preventers of innovations seem to be limited. Nevertheless, OIC also face huge challenges hindering innovation. Thus, contributions by promotors are of greatest importance to achieve high quality innovative outcomes. These challenges include e.g. identifying the right community members for different tasks, organizing a decentralized development process or balancing different opinions among a large number of participants. The characteristics of these challenges indicate a high demand for process or relationship promotors within OIC, respectively boundary spanners.

---

[32] So-called personal union (Hauschildt and Kirchmann 1999; Witte 1999)
[33] Following Hauschildt and Schewe 1999

48

3.2   Research hypotheses

In order to fill the research gap outlined in chapter 2.3 one main research hypothesis for each of the three research questions are derived. In the further proceeding these are operationalized to enable statistical test. As pointed out in the literature review of member behavior in OIC, the empirical findings of other authors about community member specialization are limited. Therefore, the predicted community member behavior is initially deduced by applying the identified influencing factors and the theories of individual behavior outlined in the previous chapters. Afterwards the empirical findings of research scholars to related behavior categories are used to support the deduced predicted member behavior. On the one hand this approach secures the generalizability of the research hypotheses and on the other hand provides an initial explanation for the predicted individual behavior. The following chart gives an overview of the employed theories and influencing factors for each of the research hypotheses:

			Research hypotheses		
			Hypothesis 1: Contribution focus	Hypothesis 2: Contribution activity	Hypothesis 3: Knowledge transfer
Theories / Influencing factors		Choice Theory	x	x	
	Social cognitive theory	Motivation	x	x	
		Community characteristics		x	
		Knowledge	x		x
		Roles and authority			x
	Key individuals in innovation			x	x

Figure 7: Theories and factors employed for deriving the research hypotheses

Referring to the RCT, community members are utility-maximizing individuals (Besanko and Braeutigam 2007). The community members are expected to focus their contributions to maximize their perceived benefit and simultaneously minimize their costs. Following the social cognitive theory and the findings from recent literature, personal motivation can provide a substantial share for predicting how community members behave and therefore how they focus their contributions. Nevertheless, the member's knowledge and/ or previous experiences provide an additional share when trying to predict a member's contribution focus. Considering the utility-maximizing goal, community members are expected to focus their contributions on topics where they can utilize knowledge already available. Reflecting the knowledge sharing process in OIC, the time required for understanding the need for knowledge, acquiring the necessary knowledge and sharing it accurately with the community are equivalent to costs. More specifically to opportunity costs, since the community member is not able to spend the time on other personally rewarding activities. To minimize the required time and therefore the costs, members are expected to mostly share knowledge already in their possession with the community. However, knowledge also functions as a moderator for many personal motivations. Relying on knowledge already available limits the risk of bad quality contributions resulting into a possible reputation loss of the member. Personal motivations such as career enhancement and efficacy emphasize the thesis of utilizing knowledge already available. To enhance their career opportunities members focus on good quality contributions, which more likely come from their own expertise field. Furthermore, they will sense those contributions as more valuable to the environment and therefore the self-efficacy effect is strengthened by using knowledge in their possession. Anyhow, community members are expected to focus their contributions not only based on knowledge and/ or experience but also on their belonging to a peer group. Members seeking reputation among peers and reciprocity contribute to topics that are most likely recognized by peers. First, this behavior secures the recognition of the contribution by peers and second, the quality of assistance from peers exceeds that of other community members. The relationship of a community member to the community outcome can be seen as the conjunction of the member's knowledge and his peer-group. Hence, familiarity represents the main driving factor for the member's contribution focus. One of the most frequently mentioned personal motivations to contribute to OIC, one's own need, additionally supports the importance of familiarity. If community members want to benefit directly from their effort and contributions, they very likely focus on topics close to their familiarity with the community objective.

Hypothesis 1: Community members focus their contributions based on their familiarity with the innovative outcome.

As mentioned previously, even initial findings about the contribution focus of community members are rare. Van Krogh et al. only outline the contribution focus of community joiners and identify a specialization on topics in which the members have specialized in other projects (von Krogh et al. 2003). Therefore, the work by Nambisan and Mahr (2010) solely serves as a basis for supporting this first research hypothesis. In their study, they identify a high sense of responsibility for the community as the main driver for members to focus their contributions on topics related to the community. In contrast, a high sense of partnership with the community sponsor foster contributions to the company. Their identified coherences are fairly in line with the first research hypothesis. Members tend to allocate their contributions according to their familiarity either with the sponsor or the community.

Specialized community members strongly follow the utility-maximization paradigm proclaimed within the RCT. They tend to maximize their benefits and minimize their costs when participating in an OIC. This goal is expected to influence specialized members not only when focusing their contribution as predicted in the first research hypothesis but also while considering aligning their activity level. A major factor driving the benefit and cost balance is the importance of their specialization area within the community discussion. Considering the social cognitive theory again, personal motivations and the community characteristics as environmental factors should influence the behavior of the individual community member while trying to achieve the utility-maximization goal. Assuming an increase in topics from the specialization area of a community member, this member will get the chance of additional benefit. In this constellation, the member has more opportunities to gain reputation from peers, to assist others and get support afterwards, and to benefit directly from the community and its outcome. Simultaneously the member contributes less costly to the community as the member can build upon expertise and knowledge already available. Based on this benefit cost evaluation the community member is expected to be more active. In contrast, an increase in topics not from the specialization area of the member increases the risks for additional costs. The high number of topics with less relevance to the member raises the required time to identify opportunities to successfully contribute to the community. Additionally, the potential for benefiting from the community is limited. The great distance to peers and therefore between beneficiaries and contributors within the network weakens the chances of reciprocity. Unequal information benefit and quality lead to high value asymmetries within the group and limit knowledge sharing activities (Thorn and Connolly 1987). Considering this negative benefit cost balance, the respective community member is expected to be less active. Summarizing both variables, the number of topics specialized in and the number of topics not specialized in, need to be considered. Therefore, the importance of the specialized area represented by the share of topics from this area in the overall community discussion influences the member's contribution activity distinctly, thus its dominance within the community discussion.

Hypothesis 2: Community members align their contribution activity with the importance of their specialized area in the community discussion.

Empirical findings of other authors regarding the contribution activity of community members in general initially support the derived second research hypothesis. Bateman et al.

51

(2006) show that a high affective commitment, meaning an intensive feeling of belonging to the site or community, results in a high contribution activity. The affectivity of the community members is compulsory strongly depending on the amount of discussed topics from their specialization area. Additionally, if the specialization area of certain community members is mainly omitted in the discussion, they will feel socially depreciated or as subprime members and will reduce their contribution activity accordingly (von Hippel and von Krogh 2003).

Transferring tacit knowledge possessed by specialized individuals into explicit permanently available knowledge for the community is a key challenge for each community. As outlined previously in the phenomenological background, interaction between community members in a collaborative manner facilitates this transfer (Nonaka 1994). In this process, knowledge needs to be codified and made independent of an individual's mind or at least transferred to individuals heavily anchored within the community. Individuals featuring characteristics from process promotors and internal boundary spanners will most likely perform this task. They hold tight connections to various community members and are capable of overcoming intra-organizational barriers (Tushman 1977). These individuals will most likely master the challenge of being accepted by isolated sub-groups consisting of specialized community members. Assigning a role to a particular community member can provide him with the necessary lateral authority to be successful in doing this job. The expectations associated with a role will be incorporated by the appointed member into his identity (Turner 2003). His connection to the community is strengthened, his responsibility for the community is increased and he sees boundary spanning between different internal organizations as one of his core tasks. Thus, community members with dedicated roles are expected to secure the sustainable integration of knowledge from specialized participants into the community.

Hypothesis 3: Community members with dedicated roles integrate knowledge into the community by holding strong communication ties to specialized members and groups.

Empirical findings of other authors initially support the importance of roles in this process. Members staying with the group long term do more mundane but necessary tasks no one else do (Shah 2006). These individuals released from strong specialization bands seem perfectly suited to hold knowledge within the community permanently. Especially a clear role assignment to these members seems to motivate them in doing coordination work and keeping the community alive (Dahlander and O' Mahony 2010). Furthermore, Dahlander and Wallin (2006) show that dedicated sponsored community members grow into central positions of the network holding strong communication ties to a high number of other members. Thus, they are predestined to combine the knowledge from various specialized community members and integrate this knowledge permanently into the community.

Part B: Empirical study

Having defined the research framework for this thesis comprehensively, this part B documents the empirical study conducted for contributing to the research on member specialization in OIC. The chosen research design, data gathering and preparation process is presented, before a two-step approach consisting of a larger quantitative assessment followed by various detailed qualitative investigations is accomplished.

4 Research design and data sampling

This chapter outlines the research approach chosen for the empirical part of this thesis. First, case study research in general and the selected single case in particular are described and second, the extensive preparation of the raw data, enabling statistical assessments to find evidence for the derived research hypotheses is detailed.

4.1 Case study

As previously mentioned in the introduction of this thesis, the research method is a single case study investigating an OIC from the consumer goods industry. Before describing the selected case in depth by pointing out the history, activities, member groups, etc. of the community, case study research is explained in detail. Therefore, its applicability as well as its different characteristics are explained and the rationale behind choosing a single case study is justified. Finally, the process of the raw data collection is shown.

4.1.1 Case study approach

"The case study is a research strategy which focuses on understanding the dynamics present within single settings" (Eisenhardt 1989, p. 534). Contrary to an experiment, a study that includes manipulation of one or more characteristics, the case as object of the study occurs within a real life context (Dul and Hak 2008). The diverse application fields of case studies range from providing descriptions to testing or even generating theory (Eisenhardt 1989). Especially when asking 'how' and 'why' questions, the use of case studies (or experiments or histories) is likely to be preferred over large-scale surveys. "This is because these questions deal with operational links needing to be traced over time, rather than mere frequencies or incidence" (Yin 2009, p. 9). The most frequently cited shortcoming of the case study approach is a limited generalizability due to the small sample size. Nevertheless, "case studies are generalizable to theoretical propositions and not to populations or universes" (Yin 2009, p. 15). The goal in carrying out case studies is the analytical generalization, meaning to generalize or extend theory and not the statistical generalization through reckoning frequencies (Yin 2009). "Case studies typically combine data collection methods such as archives, interviews, questionnaires, and observations. The evidence may be qualitative (...), quantitative (...), or both" (Eisenhardt 1989, pp. 534–535).

Following Yin two dimensions define the design of a case study approach, the number of cases incorporated (single-case vs. multiple case) and the number of units analyzed (holistic vs. embedded) (Yin 2009). "The evidence from multiple cases is often considered more compelling, and the overall study is therefore regarded as being more robust" (Yin 2009, p. 46). The main motive underlying a multiple case study is to perform a successful replication. Employing different cases to show similar results is defined as literal replication, whereas showing contrary results but for predictable reasons is named theoretical replication (Yin 2009). The term multiple case study can be further distinguished into comparative, parallel single and serial single case study. In a comparative case study data from two or more cases are compared with each other. In a parallel single case study theory or more detailed propositions are tested in a number of cases without taking the result from one case into account for another case. Whereas the serial single case study takes the results from one case into account for the next one and so on (Dul and Hak 2008). Nevertheless, multiple case studies are very time-consuming and therefore often extend the resources available to one researcher (Yin 2009). Even if multiple case studies seem to be more acknowledged for delivering compelling results, single case study is an established and powerful research approach. Siggelkow (2007) underlines the persuasive power of case study research by describing the famous single case studies of the 'talking pig' and 'Phineas Gig'. Although these two case studies consist of a unique or rare case, the application fields for this research approach are various. In total, Yin notes five rationales for conducting a single case study (Yin 2009):

- The 'critical case' represented by a single case that challenges or extends an established and well formulated theory.

- The 'extreme or unique case' represented by a single case that is so unique or rare that the documentation of each case is highly valuable.

- The 'representative or typical case' represents a case with typical conditions or circumstances for a variety of other cases.

- The 'revelatory case' exists when a scientist has the unique opportunity to investigate a phenomenon previously not observable by other scientists.

54

- The 'longitudinal case' is the fifth rationale for conducting a single case study. The same case is studied at least at two different moments in time.

In order to successfully investigate all defined research questions, the community discussion needs to be studied at multiple moments in time. Consequently, a longitudinal case study and therefore a single case study is the appropriate approach. Nevertheless, to surmount shortcomings in this research approach, the selected case should additionally represent a typical case. The evidence from the findings are more robust and their generalizability can be more easily applied if the study setting can be found in a variety of other cases. The chosen community should meet multiple characteristics of OIC detailed previously in this thesis, such as collaboration, focus on an innovative outcome and electronical communication. Its different member groups have to fit into the sketched framework for community members in OIC. Further, considering the characteristics of the defined research questions and derived hypotheses, an embedded approach is required, resulting in more than one unit of analysis. Although different community member groups are the focus of this study, the overall community discussion needs to be incorporated for investigating the consequences of a changign community focus.

4.1.2 'Premium' as case study

'Premium' develops, produces, distributes and sells cola, beer, coffee and lemonade in Germany, Austria and Switzerland. In 2011, 'Premium' sold app. 400,000 bottles of cola, 100,000 bottles of beer and 50 kg of coffee beans. The company focuses on the organization, controlling and continuous improvement of each value activity[34] required to successfully sell their consumer products to customers. Therefore, they work jointly with external bottlers, forwarders, distributors and bars resulting in an extremely high share of outsourced services. An affiliated OIC, the 'Premium community', serves as the major source for new product, service or process developments as well as solution providers for business problems or organizational issues.

In a desperate situation, former customers of Afri cola, a well-established cola brand in Germany at the end of the 20th century, founded 'Premium' in November 2001. The founding members were part of an interest group that had been initiated two years before in 1999 after the acquisition of Afri cola by a competitor company. The concealing of the following recipe change accompanied with a taste difference bothered various customers culminating in an

---

[34] The terms value activity and value chain are described in detail later in this chapter.

interest group of more than 780 members. The members negotiated unsuccessfully with the new owner company to change the recipe back or at least launch a second brand with the old Afri cola taste. At the end of 2001, the founders of 'Premium' received the original Afri cola recipe and began to fill up the first 1,000 bottles at the former bottling company for private consumption. Already the second batch of 2,000 bottles was partly sold to end customers via different restaurants and bars in Hamburg, Germany. By the end of 2003, in consensus with external distributors and the gastronomy industry 'Premium' established initial organizational structures to enable a continuous delivery and production of its cola. Transparency for the customers, a fair treatment of all economically involved individuals or firms and a minimized carbon footprint were defined as common goals of 'Premium'. The product itself should still be the focus of further developments but the way the product is produced, distributed or sold and the kind of economic management became more important. In 2004, the rollout to other German cities beyond Hamburg started driven by different community members and today 'Premium' is still expanding its distribution network in Germany, Austria and Switzerland. In 2007 the organizational structures were professionalized embracing e.g. accounting, logistics or tax advice in order to prepare the expansion of the activities to markets beyond the cola market. Additionally 'Premium' cumulated its kind of economic management, their services and processes into an 'Open Franchise' model. Other firms or individuals can apply these process, service and management developments to create and introduce their own products in the same way 'Premium' has done. Within the year 2008 beer was introduced and 'Premium' started to fill up and deliver different cola brands as a franchise company. In 2011 whole bean coffee and in 2012 lemonade from 'Premium' entered the market successfully.

'Premium' guides all value activities required in the consumer product industry. According to Porter, all physically or technologically distinct activities performed by a company are value activities (Porter 1998). These activities can further be divided into primary and support activities. The sum of the primary activities is commonly called a value chain. It describes "the full range of activities which are required to bring a product or service from conception, through the intermediary phases of production, delivery to final customer, and final disposal after use" (Kaplinsky 2000, p. 121). The support value activities only support the primary activities by providing e.g. firm-wide infrastructure or human resources (Porter 1998). Even though the main steps of the generic value chain can be identified for each company, a refinement on industry and company level is required (Porter 1998). Referring to Müller-Stewens et al. (2005) the value activities of 'Premium' can be defined as following:

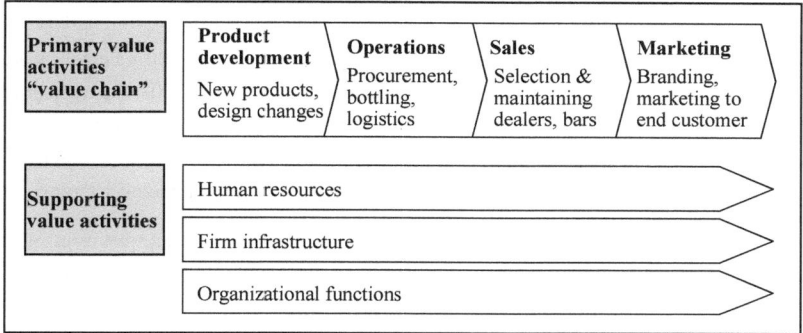

Primary value activities "value chain"	Product development New products, design changes	Operations Procurement, bottling, logistics	Sales Selection & maintaining dealers, bars	Marketing Branding, marketing to end customer

Supporting value activities	Human resources
	Firm infrastructure
	Organizational functions

Figure 8: Differentiation of Premium activities into value and supporting[35]

The affiliated 'Premium community' has escorted the activities almost from the first beginnings in 2001 and has a significant impact on 'Premium'. Community members arise from the external service providers, such as bottlers and distributors, 'Premium' representatives and end customers. All members have access to the same mailing list, which represents the main communication channel of the community. Usually the central organizer of 'Premium', occasionally other representatives and rarely other community members post problem statements about all the different issues concerning 'Premium'. Afterwards the community discusses the problems and tries to find solutions for them, ranging from new product development to marketing measures and from reorganizational to financial issues in a collaborative manner. The favored outcome of this process is of course to identify the best solution for the posted problem statement but sometimes a universally valid rule called module, which describes how to solve such problems in the future, can be additionally developed. Some examples of these universally valid rules are:

- The distance between cities and bottlers cannot exceed 600 kilometers.
- The prices are fixed, no rebates are allowed but no interests are charged either.
- Production failures are directly posted on the 'Premium' homepage.
- The only discount given applies to small order batches to compensate for increased transportation costs ('Anti-Quantity-Rebate').

Over recent years, the 'Premium community' has additionally helped to develop different new products. This embraces a lot more than the already mentioned different beverages. These developments range from different box and bottle sizes to design changes of the bottles and

---

[35] Own illustration following Müller-Stewens and Lechner 2005

57

labels.

In April 2010, representing the beginning of this study, approximately 220 members were registered on the mailing list of the 'Premium community'. At the end of 2008, the number of community members exceeded 800, but 'Premium' decided to force all members to re-register in order to reduce the list to members with high contribution willingness. The active[36] members of the 'Premium community' can be classified into five different member groups:

Central organizer: One of the co-founders, who possesses the bank account, makes most of the problem statements and is involved in almost all activities within the 'Premium' network. He is the only member working full-time for 'Premium'.

Coordinators: The coordinators represent a relatively small group of community members. They are responsible for certain corporate functions of 'Premium' such as IT, accounting or other organizational tasks. Their contribution to the 'Premium community' is highly appreciated but not mandatory to fill out their position within 'Premium'.

Commercials: This community member group consists of employees from different commercial actors of the 'Premium' network, e.g. external bottlers, forwarders, distributors and bars serving 'Premium' products. They deal with the products almost every day but for all of them 'Premium' products account for only a small share of their revenues.

Privates: This group exclusively contains end users, respectively customers, buying 'Premium' products for their own consumption.

Micro-entrepreneurs: These community members are a sub-group of the Privates group. They are end users but additionally responsible for the acquisition and maintaining of bars, dealers and shops selling 'Premium' products. Each city in Germany, Austria or Switzerland, in which 'Premium' is sold is exclusively assigned to one individual of this member group. Commonly these sales regions are initially developed by the respective Micro-entrepreneur.

The product-relation of these three member groups differs. The Commercials handle the production and logistics of the products as well as their sales. The Privates are the primary target of the 'Premium' branding and marketing measures and therefore marketing represents their contact point with 'Premium'. The Micro-entrepreneurs are also end customers like the Privates and therefore directly influenced by the marketing activities but they are additionally in charge of selling the products to dealers and bars. The following chart summarizes this differentiation of their value chain positions:

---

[36] Considering the research objectives, the investigation of the contribution focus of different community members, only active members are in the scope of the study.

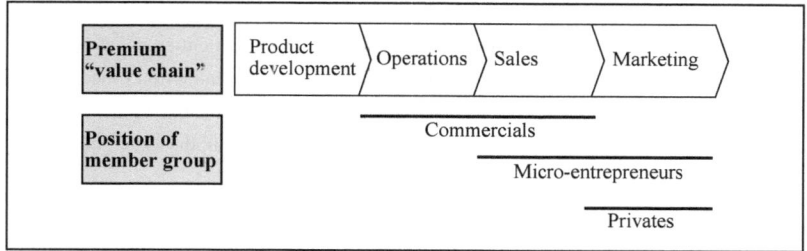

Figure 9: Differentiation of community members by their value chain position

The 'Premium community' is selected for several reasons. The community has been established for almost ten years. In this time the community members successfully proved their major importance for the development and improvement of products, services and processes. The community features a great variety of community member groups, which can be clearly distinguished from each other. In addition members with clearly assigned roles exist, the group of Coordinators. Moreover, as shown later in this thesis, a longitudinal and complete data set is available. This enables the author of this study to address all defined research questions. Nevertheless, as outlined previously, the 'Premium community' should represent a typical case to increase the evidence of the empirical findings. Therefore, the community itself and the different member groups should meet multiple characteristics of other OIC and their members.

The 'Premium community' clearly meets the characteristics describing the concept of online communities (Blanchard and Horan 1998; Rheingold 2000; Kozinets 1999; Ridings et al. 2002; Tietz 2007). First, all community members participate voluntarily, exchange content and communicate over a long period. Second, the communication is primarily electronically. Third, the individuals share a common goal and a community culture has been established. In addition, following Raasch et al. (2009) the 'Premium community' shows important characteristics of an OIC. The members contribute to a joint development of one or several designs[37] and these designs are exploited. Even though 'Premium' considers its franchise model as open, the comparison with the characteristics of openness defined by Balka et al. (2010) indicates only semi-openness. The accessibility of the 'Premium community' itself is partly restricted. New members need to know at least one member registered already to get access to the mailing list. In addition the recipe of the cola is closed and cannot be changed by a community member, but all other processes, services and management rules are accessible by all community members. Because of the transparency 'most of 'Premium's' knowledge is

---

[37] In this case different products, services and processes.

publicly available on the internet but some critical information is only shared with the 'Premium community' that is not accessible for everyone as mentioned above. The replicability of the products is obviously limited due to the complex production processes and the variety of ingredients for each product. The replicability of the franchise model, meaning the adoption of the processes and services for other products, is on the other hand fully replicable.

Besides the community itself, the members of the 'Premium community' undoubtedly show close similarities to members of other OIC. This is especially emphasized when considering Ye et al.'s (2005) definition of member roles in OSS communities and its adoption on OIC in general (van Oost et al. 2008). The central organizer is representative for the project leader and the Coordinators fulfill similar roles as the core members of other communities. The Commercials, Privates and Micro-entrepreneurs represent the developers or bug fixers and reporters. Passive users or readers are condensed within the group of community members not contributing to the 'Premium' list by sending mails. Nevertheless, these non-contributing members could also be bug reporters since the process of bug reporting differs slightly from other communities. Bugs are frequently reported directly to the project leader and he posts these issues instead of the bug identifier. The differentiation of the 'normal' developers or bug fixers or reporters into three groups also follows generally applicable characteristics. Community members are often classified into Commercials and Privates, (e.g. von Hippel 2007; Balka et al. 2009) and the Micro-entrepreneurs are analogous to free lancers or self-employees identified by fellow researchers (Hars and Ou 2001; Ghosh et al. 2002).

4.1.3   Raw data collection

Considering the defined research questions and hypotheses, the contributions to the 'Premium community' by different members represents the research unit. These contributions consist of the individual mails sent to the mailing list. Communication data prior to March 2005 is not available and therefore all mails from the list starting in March 2005 until April 2010 represent the raw data for further proceeding. This is a 62-month period with 7,362 mails from 180 different mail accounts. The author of this thesis registered to the list in April 2010 in order to get comprehensive access to the required data, but did not contribute to the list.

4.2   Data preparation

In order to assess the contribution focus, the contribution activity and the communication between different community members, the raw data consisting of several thousand mails need to be prepared accordingly. Therefore, the author conducts a detailed content analysis of all contributions to the mailing list. Two recruited research scholars who independently from each other code the contributions from a smaller sample validate the assessments. Afterwards the co-citation analysis emerging from bibliometric studies is applied in an innovative manner to identify communication ties between different individuals and member groups of the 'Premium community'.

4.2.1   Content analysis

Content analysis can be defined as a "systematic, objective, quantitative analysis of message characteristics" (Neuendorf 2002, p. 1). Analyzing the content of customer postings or mails from online forums or mailing lists has been employed in various studies of OIC (e.g. Wasko and Faraj 2005; Füller et al. 2007). Following Krippendorf (2004) a three-step approach consisting of sampling, unitizing and coding is applied in this study. Starting from the 7,362 mails from the mailing list 134 mails are excluded to generate the sample to be unitized. The mails excluded were sent either by externals not identified as community members or by registered community members with a total mail amount of one. These solitary mails often contain the request to unsubscribe the respective member from the mailing list and therefore are immaterial for the community communication. The remaining 7,228 mails from 105 different community members including the central organizer are grouped into 1,371 discussion threads, which serve as a basis for the main part of each content analysis, the coding procedure.

4.2.1.1   Coding procedure

In a very simple way, coding is defined as "the process of analyzing data" (Strauss and Corbin 1991, p. 61). The coding procedure is essential since "(...) content analysts need to transform unedited texts (...) into analyzable representations" (Krippendorff 2004, p. 84). Within content analysis, this process contains the conceptual labeling of e.g. discrete happenings and a classification of these concepts into different categories. While conducting this classification "(...) concepts are compared one against another and appear to pertain to a similar phenomenon. Thus the concepts are grouped together under a higher order, a more abstract concept called a category" (Strauss and Corbin 1991, p. 61). On the one hand, the categories, respectively the coding scheme, can emerge inductively during the interaction with the data and are continuously refined throughout the entire analysis process (Hara and Hew 2007). On

61

the other hand, a deductive approach is also quite common using predetermined categories defined prior starting with the coding procedure (Nambisan and Baron 2010). The inductive, incremental process is often used in theory-building research whereas theory-testing research facilitates the utilization of the deductive approach (Elo and Kyngäs 2008). Returning to the research subject, the communication within the 'Premium' community, an approach with predetermined categories is chosen. The research hypotheses are clearly formulated, derived from individual behavior theories, and need to be supported by the community communication data. Therefore, seven different categories have been chosen, which divide the contribution content on a sufficiently detailed level. Besides the four value chain steps, Product Development, Operations, Sales and Marketing, all support value activities are combined into the fifth category Administration. This includes all different corporate functions not assigned to one primary value activity, such as human resources, IT or finance. Only the support activity Corporate Social Responsibility[38] is defined as a separate category as it presents a core asset of 'Premium' and its community and the number of contributions is predicted to be very high. The last and seventh category, Others combines all mails, which cannot be assigned to the other six categories, such as off-topics or information about the consumer industry.

After identifying the deductive approach as appropriate and the predetermination of seven categories, the content analysis ensues by coding the units to these categories. Before coding all 1,371 units, in addition to the author, two research scholars code a smaller sample of 276 units representing approximately 65 % of all mails. Subsequently the intercoder reliability between the three researchers is measured. This process is required to independently validate the assessments made by the author during the coding procedure. Due to its importance this process is thoroughly explained in the following chapter 4.2.1.2. On the strength of the intercoder reliability index and the successful resolving of the main disagreements, the remaining 1,095 discussion threads accounting for approximately 35 % of all mails are only coded by the author of this study. This approach follows recommendations and studies of other researchers[39]. To give the reader of this thesis a brief understanding of the problems solved by the 'Premium' community and the outcome of the coding procedure some examples of each category are given in the following table:

---

[38] Numerous definitions for Corporate Social Responsibility exist in today's literature. All of them consist of one or more of the five dimensions: Environmental, social, economic, stakeholder or voluntariness (Dahlsrud 2008). Moir points (2001) out a definition by Volkswagen that shows close similarities to 'Premium's' view: CSR is "the ability of a company to incorporate its responsibility to society to develop solutions for economic and social problems" (Moir 2001, p. 21).

[39] E.g. Hara and Hew 2007, Wasko and Faraj 2005, Nambisan and Baron 2010

Category	Examples from mailing list
Product Development	How could Premium develop a child-friendly version of its cola? Should Premium introduce a beer? A new bottle design is necessary due to empties problems: Which design should it be?
Operations	How to solve the problem with unsorted empties boxes? What is the most efficient way to integrate retailer XY into the established transportation routes? Where to find an alternative bottler as backup solution? How to improve the goods ordering system of some retailers? Who knows a reliable source of organic sugar?
Sales	Should Premium use a gastronomy consultant as new sales channel? Is Denmark a potential foreign market for selling Premium products, which cities are most promising? How to react to agressive sales strategy from other cola companies? Sales in city XY are declining, how to increase sales in this city?
Marketing	Is magazine XY the right medium to launch a report about Premium and its community? Call for ideas for the Premium homepage to be developed within the next months. Should Premium develop a long term marketing strategy and how should it look? Donate Premium products as marketing measure?
Administration	Which IT solutions could be established to increase the community member activity? How to deal with a retailer owing Premium money? How to allocate daily organizational tasks more efficiently to Coordinators? What is the appropiate salary for the accounting Coordinator? Which open source software tool could increase the overall efficiency of the Premium network?
Corporate Social Responsibility	Are contracts guaranteeing exclusiveness socially correct? How to use the "Alkohol-Cent" from each bottle of beer sold? Is company XY the right one to compensate CO_2? Should Premium introduce the ample logic voluntarily?
Others	Post about interesting report of the cola market in Germany. Link to a funny commercial for Fanta. Should Premium use the female or male version when communicating?

Table 3: Extract of discussion threads from the 'Premium' community

After successfully coding all units, the data preparation is continued by adding the month and year each thread was discussed and the mails sent by each community member to the respective thread. At the end, one community member is excluded in order to finalize the data preparation. After discussion with the central organizer of 'Premium', this member habits a special status within the 'Premium community'. This member is the only co-founder still active (besides the central organizer) and its activity outranges all other members by almost 100%. To prevent any distortion of the results caused by this one individual, the member is dismissed from the final data set of the content analysis. Someone might argue to additionally exclude the central organizer from the data set as he also habits a co-founder status. Nevertheless, as he is not included in further analyses his data is kept in the final data set. That is shown in the following table serves as a basis for all further analysis. It represents the quantification of the community communication through the mailing list and its structure follows suggestions by other authors, e.g. Krippendorf (2004):

Thread	Date		Category							Mails from community member				
Number	Month	Year	Product Development	Operations	Sales	Marketing	Corporate Social Responsibility	Administration	Others	Organizator	Member 1	Member 2	...	Member 103
1	3	05				x				3	2			5
2	3	05	x							2				2
3	3	05						x			2	1		
...														
1376	4	10			x					4	1			

Figure 10: Quantification of the communication data (Result content analysis)

4.2.1.2 Intercoder reliability

Two aspects are of paramount importance for the success of this process. The selection of the individuals conducting the coding called 'coders' and the appropriate index to measure the reliability of the coding procedure. Besides other qualities, such as necessary cognitive abilities, the appropriate background of the coders involved is most important (Krippendorff 2004). The coders should be familiar with the phenomenon under consideration and the reliability is significantly increased when the "content analysts have similar backgrounds and so (...) interpret the written constructions alike" (Krippendorff 2004, p. 128). This involves not only the professional but also the educational and cultural background (Peter and Lauf 2003). Taking the predetermined categories and the characteristic of the units into account, the potential coders require a deep understanding of corporate functions and the value chain as well as experience in online communication between innovative users. Three persons, the author of this thesis and two research scholars, have been identified as qualified for coding the unitized threads. All three coders hold a degree in business administration, have worked for a management consultancy for several years and are currently doing research on phenomena from the user and community innovation field. The reliability of the data generated by the coders during the coding procedure is paramount "since the researcher's subjectivity must be minimized to obtain a systematic, objective description of the communications content" (Kassrjian 1977, p. 13). The reliability can be manifested in various ways, but all are functions of the agreement achieved between coders, observers or judges (Krippendorff 2004). The agreement among these individuals can be measured applying the intercoder reliability. The intercoder reliability is "the degree of consistency between coders applying the same set of categories to the same content" (Kassrjian 1977, p. 14). After coding the data or a sample, one index or more indices of reliability are typically evaluated and reported. Despite manifold indices exist only a handful of techniques are used frequently (Lombard et al. 2002). Minimizing the reader's efforts only intercoder reliabilities indices applicable for this case and data set are explained in the following.

The simplest agreement, the percentage agreement "is the percentage of all coding decisions made by pairs of coders on which the coders agree" (Lombard et al. 2002, p. 590). The main advantage of this index for intercoder reliability is obviously its straightforwardness but it does not involve the agreement among coders simply by chance. Especially in coding procedures with a small number of categories, the chance of occasional agreement among the coders is not negligible. Therefore, most of the studies based on content analysis report at least one additional intercoder reliability index. Considering the characteristics of the applied coding approach, employing multiple coders and a nominal scale, Fleiss Kappa is the appropriated index to calculate the intercoder reliability. Fleiss Kappa is derived from the widely used Cohen's Kappa[40]. In contrast to the percent agreement, Cohen incorporates the agreement by chance probability into his agreement coefficient. He states that "for any problem in nominal scale agreement between two judges, there are only two relevant quantities" (Cohen 1960, p. 39):

p_o = the proportion of units in which judges agreed

p_c = the proportion of units for which agreement is expected by chance

"The coefficient Kappa is simply the proportion of agreement after chance agreement is removed from consideration:" (Cohen 1960, p. 40)

---

[40] E.g. Wasko and Faraj 2005; Nambisan and Baron 2010

$$\kappa = \frac{p_o - p_c}{1 - p_c}$$

The value of Cohen's Kappa ranges between -1.00 and +1.00. More agreement than by chance results into a positive value. Whereas less agreement than by chance vice versa results into a negative value. Cohen's Kappa represents a conservative index to measure intercoder reliability and therefore more liberal criteria for assessing its value are usually used (Lombard et al. 2002). Nevertheless, universally applicable minimum value does not exist, but Landis and Koch in accordance with other authors see an almost perfect agreement for values exceeding 0,81 (Landis, Koch 1977). Three assumptions are proposed to use Cohen's Kappa as a measure for intercoder reliability: "The units are independent, the categories of the nominal scale are independent, mutually exclusive, and exhaustive (and) the judges operate independently" (Cohen 1960, p. 38). To remove the restrictions of Cohen's Kappa, only two raters and both rating each subject, Fleiss extends the coefficient to cases with multiple raters and where each subject is not necessarily rated by each rater (Fleiss 1971):

$$\kappa = \frac{\overline{P} - \overline{P}_e}{1 - \overline{P}_e}$$

with $\overline{P} = \frac{1}{Nn(n-1)}(\sum_{i=1}^{N}\sum_{j=1}^{k} n_{ij}^{2} - Nn)$

with $\overline{P}_e = \sum_{j=1}^{k} p_j^{2}$ and $p_j = \frac{1}{Nn}\sum_{i=1}^{N} n_{ij}$

and $n = $ *Number of raters* , $N = $ *Number of units* , $k = $ *Number of categories*

In order to reduce the coding effort, a smaller sample derived from the overall sample is used to determine the intercoder reliability. In this case the adequate size of the smaller sample needs to be determined to guarantee a desired level of statistical significance for the intercoder reliability index. Therefore, a formula from Bloch and Kraemer (1989) is applied, which implies for the required minimum number of units N:

$$N = z^2 \left(\frac{(1+\alpha_{\min})(3-\alpha_{\min})}{4(1-\alpha_{\min})p_c(1-p_c)} - \alpha_{\min} \right)$$

The required sample size is a function of the minimum acceptable reliability $\alpha_{\min}$, the smallest estimated proportion of units from a certain category p_c and the corresponding z value for the one-tailed test. If an estimation of the expected smallest proportion of units prior to the coding procedure seems imprecise, assuming an equal likelihood of each category but adding units to the sample after identifying unequal proportions in the data is an acceptable method

66

(Krippendorff 2004). Referring to the predetermined seven categories an equal distribution of the units results into a proportion of 0.143. Considering that some categories include more subjects, e.g. Administration, than others, an unequal distribution seems to be very likely. Therefore it is assumed, that the smallest expected proportion is equivalent to half the proportion for an equal distribution resulting in $p_c= 0.072$. Even though Landis and Koch suggest 0.81 as an almost perfect agreement, in this study $\alpha_{min}= 0.85$ is used for the determination of the smallest sample size in order to be in accordance with other studies. Together with z= 1.645, representing a 5 % statistical significance level (Bortz 2005), the number of units for the intercoder reliability sample should be at least 267. For the final generation of the sample to be coded by all three coders, the discussion threads are sorted descending by the number of mails sent to the respective thread. This guarantees that the most important threads with respect to the community communication intensity are coded by all three scholars. Finally, 276 threads are included in the intercoder reliability sample including all threads with at least eight mails. Therefore, the generated sample contains 20 % of all discussion threads and approximately 65 % of all mails.

Afterwards this sample is studied in detail and independent of each other by the three coders and each thread is finally allocated to one of the seven categories. The three independent coders achieve a total agreement in 233 of the 276 threads, which represents a percentage agreement of 84%. The percentage agreement rises strongly if only an agreement of at least two coders is considered. In this case, an agreement in 270 of the 276 threads is achieved, resulting in a percentage agreement of 98%. Since the percent agreement is a very simple method of measuring intercoder reliability with several weaknesses, Fleiss Kappa is applied as an additional agreement index. The calculated value of 0.87 represents high inter-coder reliability, especially when considering the high number of categories used in this coding process, since an unweighted Kappa tends to be lower with an increasing number of categories (Maclure and Willett 1987). Additionally it shows accordance with the published work of other researchers[41]. The smallest proportion of one category is 0.082 for Others. This verifies the number of units (276) included in the intercoder reliability sample since 0.072 was the estimated smallest proportion of one category when determining the minimum sample size of 267. By reviewing each disagreement in an in-depth discussion, the three coders identify two major failure sources. The first one, a slightly different understanding of the Product Development category, is resolved by agreeing on a common definition of the value chain step Product Development. Threads concerning cost calculation were shifted to the category Administration. Especially within the consumer goods sector pricing can be seen as some kind of product development. However, in this case the community does not run an

---

[41] Cf. 0,84 (Nambisan and, Baron 2010) and 0,86 (Wasko and Faraj 2005)

active pricing approach but rather discusses how to reward all parties involved in this venture equitably. The second failure source, difficulties to differentiate exactly between the non-value-chain topics Corporate Social Responsibility, Administration and Others, is of minor importance for this study since only the value-chain categories are finally incorporated into the statistical analyses [42]. If no total agreement is achieved by all three coders the category that is favored by two coders is chosen. After reviewing 137 research studies based on content analysis, Lombard et al. suggest guidelines for reporting the intercoder reliability (Lombard et al. 2002). According to their guidelines, the report of the intercoder reliability in this study matches all requirements that are applicable for this case. The size of the reliability sample and its relationship to the full sample are documented. Information about the number of coders, if the researcher is included and which amount of coding was explicitly conducted by which coder is given. The selection of the appropriate index is justified, the disagreements are discussed and resolved in the full sample and further information about the intercoder reliability can be requested at the author of this study.

4.2.2 Co-citation analysis

Co-citation analysis is classed among the bibliometric techniques. These techniques are applied to track objective indicators of research scholar activities, meaning mostly publications and citations. They can also be used for other processes, inputs and outputs associated with science such as scientific manpower, journals or federal funding patterns (Cronin 2001). In recent years, co-citation analysis has been well established "for the empirical study of the structures and developments of scientific communities and areas" (Gmür 2003, p. 27). It serves as a basis for identifying 'invisible colleges', representing a network of researchers with a certain relationship but without any formal organizational bond. After reviewing various studies about this phenomenon, Lievrouw (1989) defines an 'invisible college' as "a set of informal communication relations among scientists or other scholars who share a specific common interest or goal" (Lievrouw 1989, p. 622). The starting point of each co-citation analysis is presented by a certain set of publications and their bibliographies, containing the units to be analyzed. "A co-citation is taken to exist if two references or authors appear in the same bibliography. It is interpreted as the measure for similarity of content of the two references or authors" (Gmür 2003, p. 27). As anticipated from this definition, two different approaches for analyzing co-citations exist, which are based

---

[42] E.g. a thread about how to use the "CO_2-cent", meaning that one cent of each bottle sold is used to compensate the emitted CO_2 pollution during the production and distribution process. One coder interprets this as Corporate Social Responsibility due to the sustainability concerns represented by this discussion. The other one identifies this as a finance discussion and allocates the thread to the Administration category.

upon author co-citation or document co-citation. Within the author co-citation approach the writings by one author are the units of analysis. It is based on the premise that co-citations between writings of two authors show the interrelationship of these two individuals within the research field. Authors whose writings are frequently cited together can be generally seen as related in some kind of way whereas authors who are almost never cited together are positioned relatively far apart from each other (White and Griffith 1981). When applying the document co-citation approach, publications in peer-reviewed journals are seen as the most valid indicator for the structure of a research field. Contrary to the author approach, co-citations, hence the proximity between particular references, are measured (Small 1977).

In addition to these two approaches and the resulting different units of analysis a variety of methods for conducting co-citation analyses appear within research literature. However, by evaluating six[43] different co-citation methods Gmür identifies the CoCit-Score method as the only one without any obvious restrictions. The method overrates neither the most nor the less cited authors or references and additionally prevents an overrating of co-citations between commonly cited references (Gmür 2003). The CoCit-Score between author or reference A and author or reference B, which takes values between 0 and 1, is defined as following (Gmür 2003):

$$CoCit_{AB} = \frac{(co-citation_{AB})^2}{\min(citation_A;citation_B) \; x \; mean(citation_A;citation_B)}$$

Although co-citation analyses are only used to research science disciplines based on bibiliometric data, they seem to be quite suitable for studying the communication within online communities. As previously shown many research scholars apply this method when analyzing scientific communities searching for relationships between individuals not linked by formal organization ties. This goal is achieved by exploiting the "writings of a person and not the person himself" (White and Griffith 1981, p. 163). These characteristics of author co-citation studies can be transferred when analyzing discussions of an online community. Similar to researchers from different universities and countries contributing to a scientific community by publishing their work, community members contribute to a decentralized, formally unlinked online community by e.g. revealing their information or knowledge. The threads discussed and the respective participants within these threads represent the set of publications and its bibliography within one science field. Considering the definition of co-citation, the simultaneous citation of two authors within one publication, a co-citation between two community members exists if each of these two sent at least one mail to the

---

[43] Co-citation maximum, co-citation minimum, co-citation mean, CoCit-Score, Pearson's correlation and factor analysis

same discussion thread. However, contrary to the bibiliometric co-citation this relationship indicates a communication between the two individuals. This implies that a high CoCit-Score between two individuals signals a tight communication tie and intensive knowledge exchange. The co-citation analysis and more specifically the CoCit-Score facilitate the identification of real knowledge brokers and boundary spanners within an online community. Furthermore, tight communication ties between individuals purely based on high activity by one individual are leveled for two reasons. First, even if one community member sends multiple mails to a discussion thread, the co-citation count will still be one and second, the CoCit-Score decreases with a high activity of one individual.

The author co-citation approach naturally starts by choosing the set of publications and the authors included in the analysis. This narrowing-down process leads to an appropriate coverage of the investigated science field, secures the relevance of the analysis and enables clear interpretations of the results. Approaches to limit the number of publications are e.g. utilize journal rankings by including only documents published in high ranked journals (Lee 2011) or use publications from the journal that represents the science field most appropriate (Ramos-Rodríguez and Ruíz-Navarro 2004). Additionally, the consulting of plenty of top researchers from the field and asking them to list the most influential papers (Lee 2011) or including only the top-cited publications (Ramos-Rodríguez and Ruíz-Navarro 2004) are appropriate approaches. The selection of authors is frequently accomplished by simply including the top-cited authors from the respective science field (White and McCain 1998). Adapting this narrowing-down process to the community communication means reducing the number of discussion threads and community members included in the analysis. Based on the goal of this analysis in this study, to investigate the collaboration between various community members from different member groups, only discussion threads with at least four participants (excluding the central organization) are included in the co-citation analysis. This selection firstly enables the participation of all four member groups (Coordinators, Commercials, Micro-entrepreneurs, Privates) and secondly limits the probability of including threads with a low collaboration level. The set for the analysis finally consists of 280 threads encompassing app. 62 % of all mails sent by the community members excluding the central organizer. The lower limit for selecting the community members included is set to at least five contributions, which represents an activity of more than 10 % of the mean contribution activity. Finally, 75 community members are selected for the co-citation analysis.

Following Schäffer et al. (2006) three steps are necessary to determine the required CoCit-Score matrix after successfully restricting the analysis sample. Starting from the set of discussion threads and community members, the number of citations from each member needs to be counted. Afterwards the co-citation matrix is created by determining the number of co-citations between each community member pair. Based on these two data sets the

CoCit-Score matrix is derived using the CoCit-Score formula stated earlier in this thesis. The following figure shows the structure of the different data sets generated:

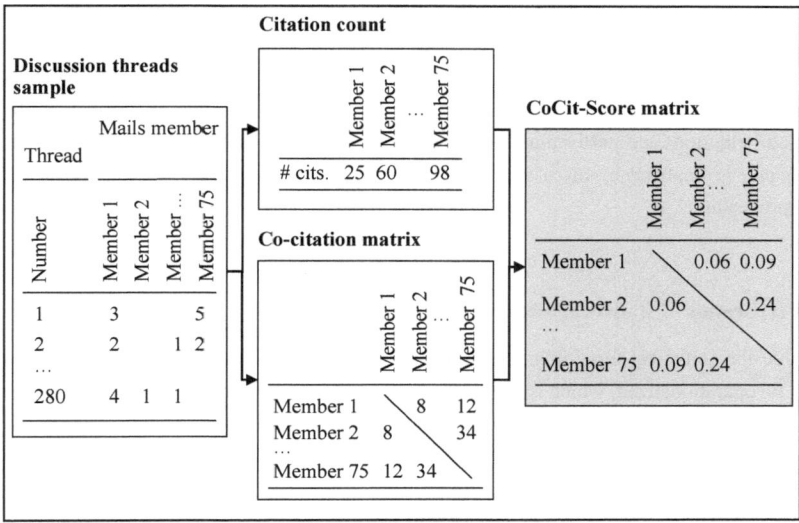

Figure 11: Generation of CoCit-Score matrix of the 'Premium community'

Normally a threshold value[44] of the CoCit-Score is employed in other studies to show sufficiently differentiated clusters (Schäffer et al. 2006). Within this study, no threshold value is applied. On the one hand, the goal of this analysis is rather to measure the communication and collaboration on a micro-level between individual community members than to identify clusters. Especially in this case, a CoCit-Score of zero helps to understand the community collaboration comprehensively as it shows no communication between two individuals at all. On the other hand, the number of included discussion threads exceeds the usual number of publications greatly[45] and therefore a lower number of co-citations between the community members is expected. This effect is additionally amplified by the fact that the majority of the analyzed threads include less than ten participants, whereas publication bibliographies rarely contain less than 20 different authors.

---

[44] Between 0.2 and 0.3, cf. e.g. Lee 2011 or Schäffer et al. 2006
[45] Frequently approx. 100 publications analyzed, cf. e.g. Ramos-Rodríguez and Ruíz-Navarro 2004

5 Quantitative analyses and results

After defining the research setup by selecting an appropriate case and conducting the detailed content and co-citation analysis, the support of the derived research hypotheses quantitatively is the next compulsory step. First the hypotheses and variables are detailed and important general community figures are discussed briefly. This is followed by a short descriptive analysis, a determination of the data distribution and finally the appropriate statistical appraisal for each of the three research questions and associated hypotheses. For evaluating the data distributions and performing the statistical tests, the software tool SPSS 17.0 is used. This chapter is finalized by discussing the quantitative results and the determination of the further proceeding.

5.1 Preparation of statistical assessments

To enable the statistical testability, the derived research hypotheses from the third chapter need to be operationalized, which is outlined in the following chapter. Additionally the model variables used in the statistical assessment and their extraction from the data are explained in detail.

5.1.1 Operationalization of research hypotheses

According to the first research hypothesis, community members focus their contributions based on their familiarity with the community outcome. As shown in Chapter 4.1.2 this familiarity with the product can be identified for three different community member groups. The product-relation of the Commercials, Micro-entrepreneurs and Privates differ depending on their value chain position. According to their positions six hypotheses can be operationalized as shown in the following figure:

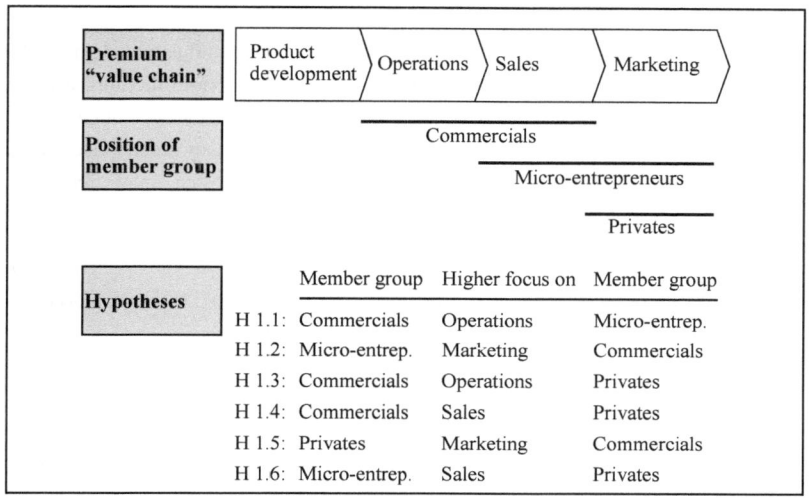

Figure 12: Operationalization of research hypotheses based on value-chain position

H 1.1: Commercials show a stronger contribution focus on Operations than Micro-entrepreneurs

H 1.2: Micro-entrepreneurs show a stronger contribution focus on Marketing than Commercials

H 1.3: Commercials show a stronger contribution focus on Operations than   Privates

H 1.4: Commercials show a stronger contribution focus on Sales than Privates

H 1.5: Privates show a stronger contribution focus on Marketing than Commercials

H 1.6: Micro-entrepreneurs show a stronger contribution focus on Sales than   Privates

The category of topics predominately discussed within a community changes over time. Especially the activity of specialized community members should be influenced significantly by this community focus. Following the second research hypothesis from the third chapter, specialized community members should increase their relative activity accordingly to an increase in the importance of their specialized category. Vice versa they should also reduce their relative activity significantly if the opposing category is posted more frequently. According to the specialized categories hypothesized for the three community member groups, eight hypotheses can be operationalized:

H 2.1: Commercials increase their relative contribution activity with an increased importance of Operations threads

H 2.2: Commercials increase their relative contribution activity with an increased importance of Sales threads

H 2.3: Commercials decrease their relative contribution activity with an increased importance of Marketing threads

H 2.4: Micro-entrepreneurs increase their relative contribution activity with an increased importance of Marketing threads

H 2.5: Micro-entrepreneurs increase their relative contribution activity with an increased importance of Sales threads

H 2.6: Micro-entrepreneurs decrease their relative contribution activity with an increased importance of Operations threads

H 2.7: Privates increase their relative contribution activity with an increased importance of Marketing threads

H 2.8: Privates decrease their relative contribution activity with an increased importance of Operations threads

Finally, the third research hypothesis emphasizes the importance of the fourth community member group, the Coordinators. Members with assigned roles hold strong communication ties to a variety of specialized members and hence secure the sustainable transfer of knowledge into the community and foster collaboration among different member groups. According to the explained applicability of the CoCit-Score for analyzing community communication, the Coordinators should show high CoCit-Scores to all other community member groups and therefore three hypotheses arise:

H 3.1: Commercials show higher CoCit-Scores to Coordinators than to any other community member group

H 3.2: Micro-entrepreneurs show higher CoCit-Scores to Coordinators than to any other community member group

H 3.3: Privates show higher CoCit-Scores to Coordinator than to any other community member group

5.1.2   Model variables

Before proceeding with the results, the variables used for the statistic assessment are explained in detail. The following chart shows the applied sources for gathering the required data to support the hypotheses:

Figure 13: Data set used in the quantitative part of the empirical study

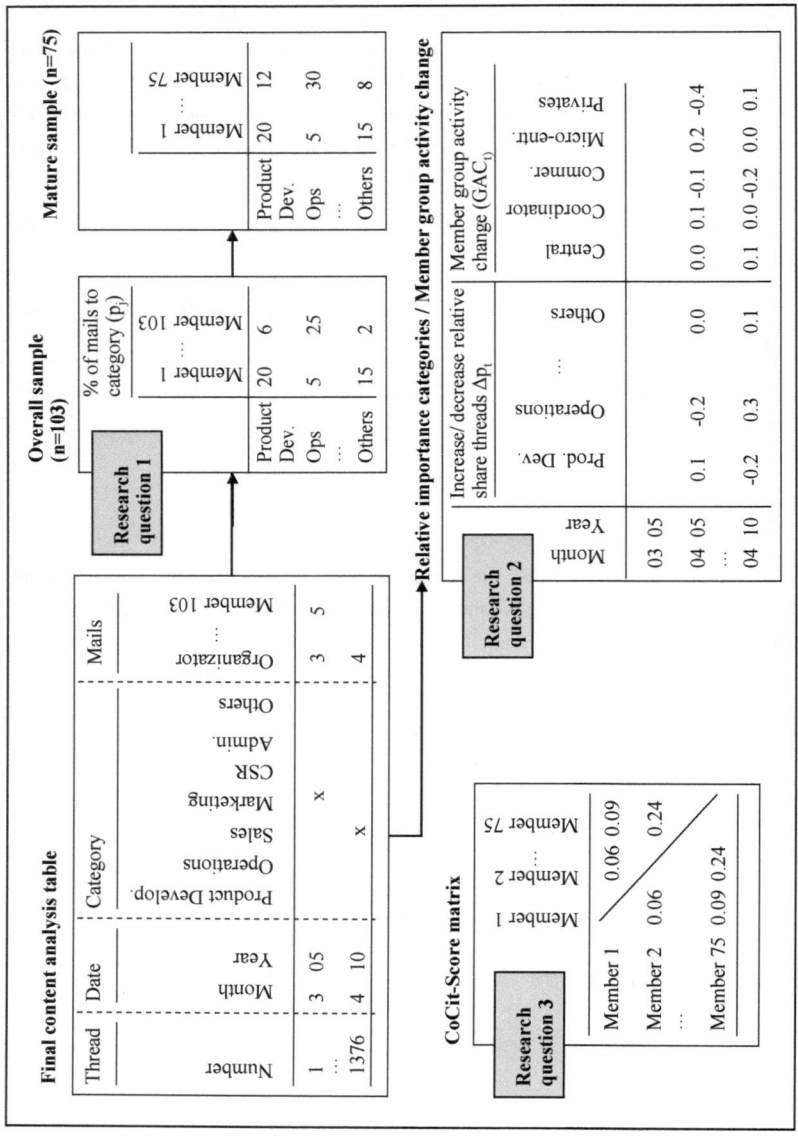

In order to support the hypotheses for the first research question, differences in the contribution focus, the specific contribution behavior for each community member is derived from the final table of the content analysis. Each member shows a specific contribution distribution characterized by the percentage of mails sent to each of the seven categories:

$$\sum_{j=1}^{7} p_{ij} = 1$$

with $p_{ij} = \dfrac{n_{ij}}{N_i}$

and n_{ij} = mails to category j by community member i,

and N_i = total number of mails by community member i.

The share of mails sent to the respective category represents the dependent variable 'p_{ij}'. By predicting a different contribution focus depending on the group affiliation of the community member, the independent variable is the *'community member group'* (Commercials, Micro-entrepreneurs or Privates) of the member i. For this cluster of hypotheses, two different samples are derived from the content analysis data. The first sample called 'Overall sample' consists of all community members included in the content analysis characterized by a contribution of at least two mails. The second sample called 'Mature sample' consists of all community members included in the co-citation analysis and therefore with a contribution activity of at least five mails. This differentiation may indicate certain specialization level changes over time. Von Krogh et al. (2003) suggest investigations of this temporal aspect after identifying the specialization phenomenon. Community members who joined the community recently may not have identified their specialization area and therefore deteriorate the results of the overall sample. Nevertheless, vice versa new members could join the community exactly in their specialization area or on easy-to-solve problems as argued by von Krogh et al. (2003).

For the second cluster of hypotheses, focusing on activity changes of specialized members, the derivation of the variables is much more complex. Starting from the final table of the content analysis again, the monthly activity change of each member needs to be initially determined. Therefore, the share of a member of the overall contribution in each month serves as a basis:

If $\dfrac{N_{i,t}}{N_t} > \dfrac{N_{i,t-1}}{N_{t-1}}$

with $N_{i,t}$ = total number of mails from community member i in month t, and $N_{i,t-1}$ = total number of mails from community member i in the last month t-1,

and N_t = total number of mails in month t,

and N_{t-1} = total number of mails in the last month t-1

the community member i increases his relative activity. Vice versa

if $\dfrac{N_{i,t}}{N_t} < \dfrac{N_{i,t-1}}{N_{t-1}}$ the community member decreases his relative activity and

if $\dfrac{N_{i,t}}{N_t} = \dfrac{N_{i,t-1}}{N_{t-1}}$ the activity level of the community member i does not change.

All three possible activity changes are integrated into the new variable Member Activity Change $MAC_{i,t}$ of the community member i in the month t. The value of this variable can either be +1, if member i has increased his activity or 0, if the member i has not changed its activity or -1, if the member i has decreased its activity in the observed month t. Therefore, this variable follows a signum function:

$$MAC_{i,t} = \operatorname{sgn}\left(\frac{N_{i,t}}{N_t} - \frac{N_{i,t-1}}{N_{t-1}}\right)$$

As previously predicted, the community members change their activity according to their affiliation to a member group, the individual MAC are aggregated to a Group Activity Change variable $GAC_{k,t}$. The index k represents the community member group and ranges from 1 for the Central Organizer to 5 for the community member group Privates. To calculate the monthly GAC for group k only MAC from the community members belonging to the group k are included (represented by the additional k index):

$$GAC_{k,t} = \frac{\sum_{i=1}^{104} MAC_{i,k,t}}{CountMAC_{i,k,t}}$$

Dividing the total MACs by the number of community members from the group k, who actually have a MAC for the respective month[46], leads to an appropriate indicator for the activity of all individual community members affiliated to one specific group. The value of the GAC ranges from -1 to +1. Whereas a negative value means that more community

---

[46] The calculation of the $MAC_{i,t}$ is obviously only possible if the community member i has made a contribution in the month t or in the month t-1.

members from group k reduced their relative activity in the month t, a positive value shows that more community members increased their relative activity in the month t.

As hypothesized, the relative importance of their specialization area should drive the contribution activity of specialized members significantly. For the month t the relative importance of a category j can be shown by the percentage share of discussed threads from category j compared with all discussed threads in the month t:

$$p_{j,t} = \frac{m_{j,t}}{M_t}$$

with $m_{j,t}$ = discussed threads of category j in the month t

and M_t = discussed threads in the month t

Assessing if a specific category is becoming more or less important in a certain month is subsequently the relative change of the percentage share compared with the previous month:

$$\Delta p_{j,t} = p_{j,t} - p_{j,t-1}$$

Like the previously derived GAC, the value for $\Delta p_{j,t}$ ranges from -1 to +1. A negative value shows a less importance of category j in month t than in month t-1. Vice versa, a positive value symbolizes a relative increase in importance of the category j. Summarizing, the independent variable for the second research question, investigating the change in the contribution activity of specialized members, is '$\Delta p_{j,t}$' (Operations, Sales, Marketing) and the dependent variable is '$GAC_{k,t}$' (Commercials, Micro-entrepreneurs, Privates).

In order to support the third set of hypotheses, investigating the knowledge transfer from specialized community members into the community, the required variables can be directly extracted from the previously derived CoCit-Score matrix. As hypothesized, each of the three specialized member groups should show higher CoCit-Scores to the group of Coordinators than to any other group. Therefore, the independent variable is the 'community member group' (Coordinators, Commercials, Micro-entrepreneurs, Privates) and the dependent variable is represented by the 'CoCit-Score' (Commercials, Micro-entrepreneurs, Privates).

5.2 General community figures

The traffic of the 'Premium' community can be best characterized by two indices: the number of mails sent to the mailing list each month but also the number of threads posted and discussed each month. The following figure shows these two indices over the periods of 62 months:

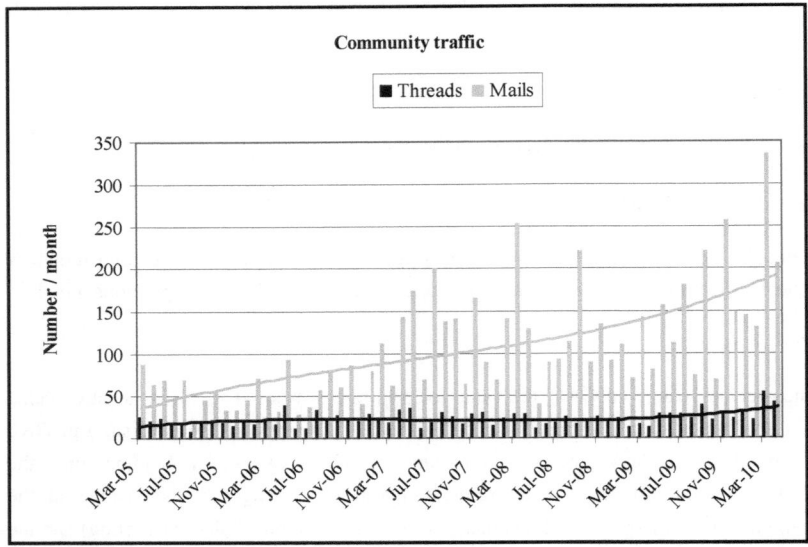

Figure 14: Community traffic from March 2005 until April 2010

The average number of threads posted per month is 22.1 and the number of mails sent 105.3 resulting in a 4.8 mails per thread ratio. Compared to the rather continuous monthly number of threads, the amount of mails is relatively less steady, which is shown by a relative high standard deviation of 64.8 (>60% of average) compared to only 8.2 (<40% of average) for the number of threads posted. Referring to the descriptive analysis of the community traffic, both the number of threads and mails increase over time. Nevertheless, the increase in mails outranges the growth in threads notably. This indicates more intensified discussions with a higher majority degree of the community. Research scholars (e.g. Schoberth et al. 2003) also observe this effect. The activity of the different community member groups is detailed in the following table:

Member group	N	Min	Max	Sum	Mean	Standard Deviation
All members	104	2	1967	6529	62.8	206.0
Excl. Central Org.	103	2	434	4562	44.3	83.3
Coordinators	10	2	434	979	97.9	151.0
Commercials	26	2	156	764	29.4	42.4
Micro-entrepreneurs	36	2	363	1528	42.4	80.9
Privates	31	2	388	1291	41.6	80.3

Table 4: Activity of different community member groups

The average activity of all community members excluding the central organizer is 44 mails. However, major differences in the activity level between the four member groups can be observed. The activity of the Coordinators outranges the other groups (average of 98 mails) by far. This finding is in accordance with studies from other researchers (Dahlander and O' Mahony 2010; Lee 2011), who also identify an increased activity of community members with dedicated roles. The activity of the other three member groups is almost balanced with a slightly lower activity level of the Commercials. This finding additionally supports the former decision to exclude the one co-founder who is still an active community member. This single Micro-entrepreneur would increase the average activity of the corresponding member group to more than 60 mails, resulting in a big activity imbalance between the three member groups that are mainly included in further analyses. The elimination of this one member also explains the difference in the total number of mails (6,529) and the number selected for the content analysis (7,228).

Another interesting descriptive analysis of the community is the distribution of all threads and mails among the different categories, as it represents the focus of the community communication. This cumulative outcome of the content analysis is presented in the following figure:

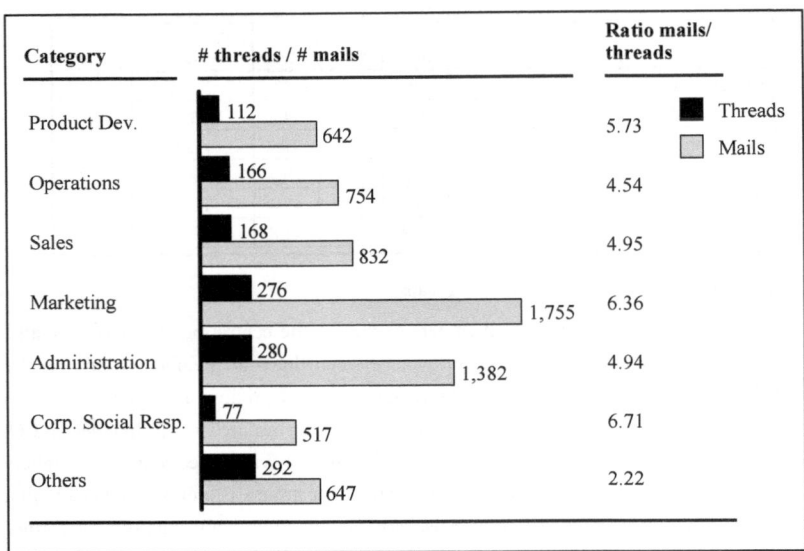

Category	# threads / # mails	Ratio mails/ threads
Product Dev.	112 / 642	5.73
Operations	166 / 754	4.54
Sales	168 / 832	4.95
Marketing	276 / 1,755	6.36
Administration	280 / 1,382	4.94
Corp. Social Resp.	77 / 517	6.71
Others	292 / 647	2.22

Figure 15: Mails and threads of 'Premium community' classified into categories

Considering the number of mails and threads, the 'Premium' community displays a strong focus on the Marketing and Administration category. These two categories combine almost 50% of all mails and approx. 40% of all threads. The mail volume to all other categories is almost balanced with the lowest volume to Corporate Social Responsibility threads. However, threads from this category are discussed most intensively (6.71 mails per thread) followed by Marketing threads (6.36 mails per thread). The category Others shows by far the lowest mails per thread ratio. This is an expected finding as these threads often include off-topics or links to other homepages and therefore feature less need for discussions. The contribution focus on a member group and individual level is analyzed in depth in the following chapter.

5.3 Community members' contribution focus

This chapter addresses the first research question:

RQ 1: How do community members focus their contributions?

Six hypotheses regarding the specialization areas of the three member groups Commercials, Micro-entrepreneurs and Privates have been operationalized at the beginning of the fifth chapter. These hypotheses are now statistically assessed using the two different generated samples: the 'Overall sample' with all 103 community members and the limited 'Mature sample' with the 75 most active community members.

5.3.1 Descriptive analysis

The next figure shows the distribution of the contributions made by the Commercials, Micro-entrepreneurs and Privates as well as the mean value of these three groups:

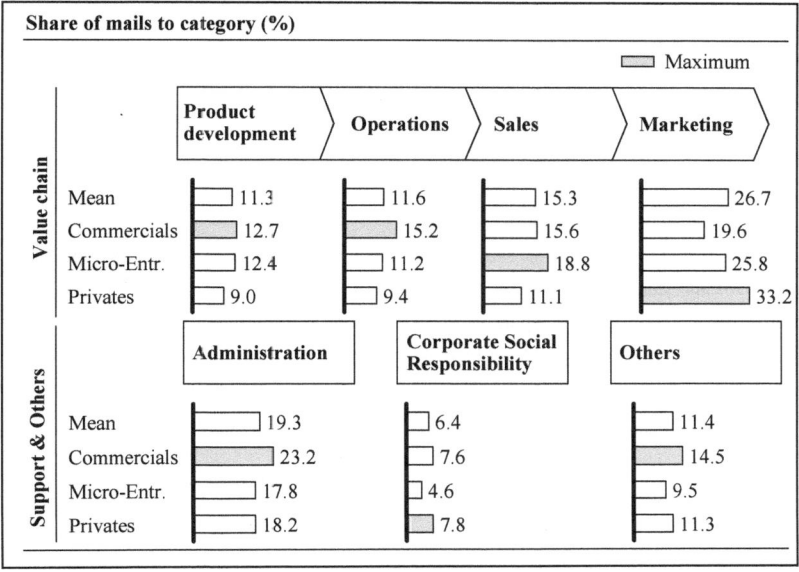

Figure 16: Share of mails to categories by member groups

An initial purely descriptive analysis of the results indicates a validation of the predicted behavior. In general, the differences between the member groups are slightly more distinct for the value chain categories than for the support categories. This effect was predicted for the categories Operations, Sales and Marketing as these represent the specialization areas of the different member groups. The Commercials show the predicted focus on the Operations category if compared to the other two member groups. Nevertheless, they cannot escape completely from the overall community focus on Marketing, which is shown by their fairly high percentage of contributions made to the Marketing category. The Micro-entrepreneurs strongly focus on Sales, whereas the Privates show the predicted strong focus on Marketing and seem to avoid Operations and Sales threads.

This initial assessment additionally provides new insights exceeding the previously derived hypotheses. The differences for the category Product Development are less distinct and probably explainable by the type of developments accomplished by the 'Premium community'. Especially the introduction of different bottle and box sizes fosters the contribution of Commercials due to handling, transport and sales issues of these newly developed products. Contrary to the other member groups, the Micro-entrepreneurs display a higher focus on the value chain activities. On average, they made approx. 68 % of their contributions to Product Development, Operations, Sales and Marketing whereas the other groups only show a share of 63 % respectively 62 %. These differences should be additionally addressed while proceeding.

However, a statistical assessment is required to confirm a sufficient significance level of the observed differences.

5.3.2 Data distribution

In order to determine the appropriate statistical test in order to verify the research hypotheses, a validation of the data distribution is required. Therefore, the shares of mails sent to the categories Product Development, Operations, Sales and Marketing by each of the three groups need to be checked on normality. The following figure shows the histograms for the 'Overall sample':

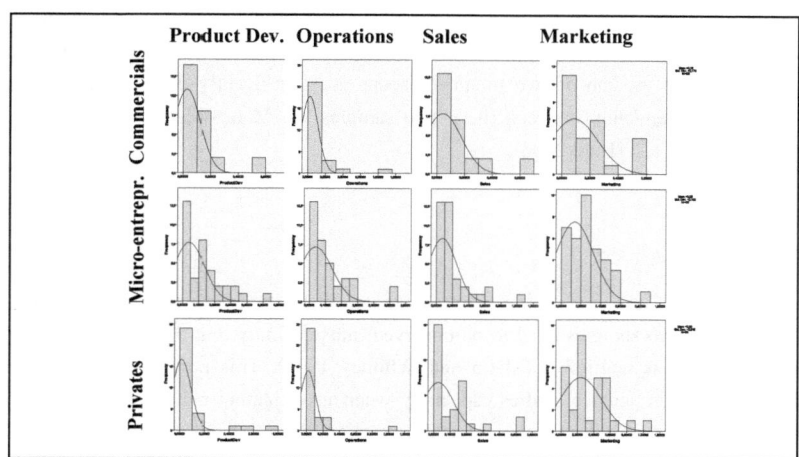

Figure 17: Histograms value chain categories and member groups ('Overall')

According to this graphical assessment, normality is rejected for all of the shares. Most of the distributions show a notably strong right skewness. The histograms for the 'Mature sample' follow accordingly:

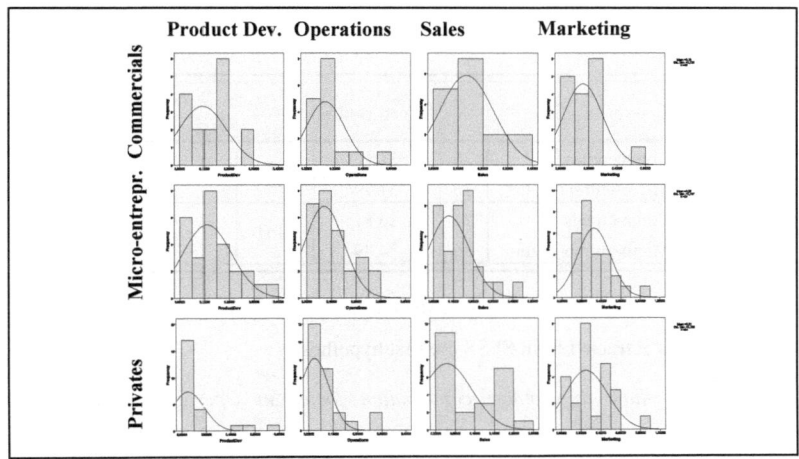

Figure 18: Histograms value chain categories and member groups ('Mature')

Even though the histograms for the 'Mature sample' show an improvement in terms of right

85

skewness for most of the variables, normality is still rejected for almost all shares. Therefore, only non-parametric tests can be applied. Considering the characteristic of the hypotheses, whether the mean or median of two member groups are significantly different from each other, and the independency between these two samples the Mann-Whitney-U test is the appropriate statistical test (Bühl 2006).

5.3.3   Mann-Whitney-U test

The Mann-Whitney-U test assesses if one of two samples is likely to have a higher value than the other one. The two samples need to be observed independently, but the characteristic of the data distribution is negligible (Mann and Whitney 1947). This non-parametric test is applied broadly within science studies especially when investigating behavioral patterns of different groups (Ruxton 2006). In the proceeding section, the results of the Mann-Whitney-U-test comparing the individual contribution distribution 'p_{ij}' for the 'overall' and 'mature' samples are outlined:

'Overall sample'

The following six tables (table 5 to table 10) show the results achieved for the 'Overall sample':

Category	Groups	N	Mean ranks	U	Significance level
Operations	Commercials	26	31.10	457.5	0.879
	Micro-entrepreneurs	36	31.79		
Marketing	Commercials	26	26.81	346.0	0.08
	Micro-entrepreneurs	36	34.89		

Table 5: Results Mann-Whitney-U Commercials vs. Micro-entrepreneurs (Overall)

Based on these results extracted from SPSS the first hypothesis:

H 1.1: Commercials show a stronger contribution focus on Operations than Micro-entrepreneurs

is not supported based on the insufficient significance level of 0.88. Additionally the mean ranks even show a slightly stronger focus of the Micro-entrepreneurs on Operations than for the Commercials. The second hypothesis:

86

H 1.2: Micro-entrepreneurs show a stronger contribution focus on Marketing than Commercials

is supported on a 0.1-significance level (α = 0.080) and the mean ranks show the predicted superior contribution focus of the Micro-entrepreneurs.

Category	Groups	N	Mean ranks	U	Significance level
All Value chain	Commercials	26	26.65	342.0	0.072
	Micro-entrepreneurs	36	35		

Table 6: Additional results Commercials vs. Micro-entrepreneurs (Overall)

Exceeding the two previously derived hypotheses and comparing the contributions by these two groups to all categories from the value-chain, the Micro-entrepreneurs show a stronger and statistically significant (α = 0.072) focus than the Commercials. This indicates a superior focus on the product itself, as they tend to contribute less to the supporting activities. This finding has already been mentioned in the descriptive part of this chapter.

Category	Groups	N	Mean ranks	U	Significance level
Operations	Commercials	26	31.30	348.5	0.368
	Privates	31	27.24		
Sales	Commercials	26	29.92	379.0	0.692
	Privates	31	28.23		
Marketing	Commercials	26	23.06	248.5	0.013
	Privates	31	33.98		

Table 7: Results Mann-Whitney-U Commercials vs. Privates (Overall)

According to the comparison of the contribution distribution of Commercials and Privates, hypotheses three and four:

H 1.3: Commercials show a stronger contribution focus on Operations than   Privates

H 1.4: Commercials show a stronger contribution focus on Sales than Privates

are not supported but the mean ranks show at least the right direction for the predicted contribution focus.

The fifth hypothesis:

H 1.5: Privates show a stronger contribution focus on Marketing than Commercials

is supported almost on a strong 0.01-significance level (α = 0.013) caused by the massive difference in the mean ranks.

Category	Groups	N	Mean ranks	U	Significance level
Product Development	Commercials	26	31.37	341.5	0.306
	Privates	31	27.02		

Table 8: Additional results Commercials vs. Privates (Overall)

An additional outcome of the Mann-Whitney-U test is the already identified preference of the Commercials towards the Product Development category, but the significance level is insufficiently high (α = 0.306) to finally support this finding statistically.

Category	Groups	N	Mean ranks	U	Significance level
Sales	Micro-entrepreneurs	36	37.10	446.5	0.154
	Privates	31	30.40		

Table 9: Results Mann-Whitney-U Micro-entrepreneurs vs. Privates (Overall)

By finally comparing the Micro-entrepreneurs and the Privates, the last and sixth hypothesis cannot be supported. Even though the Micro-entrepreneurs show the predicted preference towards the Sales category, the insufficient significance level (α = 0.154) leads to a rejection:

H 1.6: Micro-entrepreneurs show a stronger contribution focus on Sales than   Privates

Category	Groups	N	Mean ranks	U	Significance level
Product Development	Micro-entrepreneurs	36	38.72	388.0	0.029
	Privates	31	28.52		
Operations	Micro-entrepreneurs	36	37.07	447.5	0.157
	Privates	31	30.44		
All Value chain	Micro-entrepreneurs	36	36.19	479.0	0.32
	Privates	31	31.45		

Table 10: Additional results Micro-entrepreneurs vs. Privates (Overall)

The comparison of these two groups regarding the other categories uncovers some additional interesting findings. The strong preference for Operations by the Micro-entrepreneurs is not statistically confirmed on a sufficient significance level ($\alpha = 0.157$) but the difference in the mean ranks almost reaches the same level as for the Sales category. On top of that their superior contribution focus on the Product Development category compared to the contributions by Privates to this category is supported on a 0.05-significance level ($\alpha = 0.029$). These additional findings need to be discussed and clarified in detail later in this thesis.

'Mature sample'

The following six tables (table 11 to table 16) show the results achieved for the 'Mature sample':

Category	Groups	N	Mean ranks	U	Significance level
Operations	Commercials	16	24.56	175.0	0.302
	Micro-entrepreneurs	27	20.48		
Marketing	Commercials	16	14.16	90.5	0.002
	Micro-entrepreneurs	27	26.65		

Table 11: Results Mann-Whitney-U Commercials vs. Micro-entrepreneurs (Mature)

Similar to the 'Overall sample' the first hypothesis:

H 1.1: Commercials show a stronger contribution focus on Operations than Micro-entrepreneurs

is rejected based on the achieved significance level of $\alpha = 0.302$, but contrary to the previous

89

results the mean ranks show the predicted stronger focus of the Commercials on this value chain category. The difference in the mean ranks for the second hypothesis

H 1.2: Micro-entrepreneurs show a stronger contribution focus on Marketing than Commercials

rose immensely and supports the hypothesis now on a strong 0.01-significance level (α = 0.002).

Category	Groups	N	Mean ranks	U	Significance level
All Value chain	Commercials	16	17.50	144.0	0.070
	Micro-entrepreneurs	27	24.67		

Table 12: Additional results Commercials vs. Micro-entrepreneurs (Mature)

The focus on value-chain activities by the Micro-entrepreneurs is still supported on a 0.1-significance level (α = 0.070).

Category	Groups	N	Mean ranks	U	Significance level
Operations	Commercials	16	25.66	109.5	0.021
	Privates	24	17.06		
Sales	Commercials	16	24.50	128.0	0.079
	Privates	24	17.83		
Marketing	Commercials	16	15.25	108.0	0.020
	Privates	24	24		

Table 13: Results Mann-Whitney-U Commercials vs. Privates (Mature)

All hypotheses comparing the contribution foci of Commercials and Privates are supported for the 'Mature sample'. The third and fifth hypotheses:

H 1.3: Commercials show a stronger contribution focus on Operations than   Privates

H 1.5: Privates show a stronger contribution focus on Marketing than   Commercials

on a 0.05-significance level (α = 0.021 respectively α = 0.020) and the fourth hypothesis on a 0.1-significance level (α = 0.079):

H 1.4: Commercials show a stronger contribution focus on Sales than Privates

90

Category	Groups	N	Mean ranks	U	Significance level
Product Development	Commercials	16	24.25	132.0	0.100
	Privates	24	18		

Table 14: Additional results Commercials vs. Privates (Mature)

The preference previously identified in this chapter for the Product Development category by the Commercials compared to the Privates is intensified for the 'Mature sample'. The difference in the contribution focus is even supported on a 0.1-significance ($\alpha = 0.100$).

Category	Groups	N	Mean ranks	U	Significance level
Sales	Micro-entrepreneurs	27	28.63	253.0	0.177
	Privates	24	23.04		

Table 15: Results Mann-Whitney-U Micro-entrepreneurs vs. Privates (Mature)

The sixth hypothesis:

H 1.6: Micro-entrepreneurs show a stronger contribution focus on Sales than   Privates

is still rejected due to the insufficient significance level of $\alpha = 0.177$.

Category	Groups	N	Mean ranks	U	Significance level
Product Development	Micro-entrepreneurs	27	30.54	201.5	0.020
	Privates	24	20.90		
Operations	Micro-entrepreneurs	27	29.48	230.0	0.075
	Privates	24	22.08		
All Value chain	Micro-entrepreneurs	27	29.85	220.0	0.050
	Privates	24	21.67		

Table 16: Additional results Micro-entrepreneurs vs. Privates (Mature)

The newly identified preference for the Product Development category by the Micro-entrepreneurs is also supported on a 0.05-significance level ($\alpha = 0.020$). Additionally as opposed to the Privates their preference for Operations is now supported on a 0.1-significance level ($\alpha = 0.075$). The spike of the Micro-entrepreneurs on value-chain activities in total is even supported on a 0.05-significance level ($\alpha = 0.050$). Hence, this group has a stronger contribution focus on product related categories than the other two community member groups.

Based on the results for the differences in the contribution focus three interim conclusions can be drawn. First, community members seem to intensify their specialization level over time, as the 'mature sample' provides significant superior results. Second, the Commercials and Privates focus their contributions on categories close to their familiarity with the community outcome. Last, the contribution behavior of the group of Micro-entrepreneurs seems to be less driven by familiarity as they contribute to a greater variety of categories compared to the other groups.

5.4 Impact of community focus on contribution activity

The following chapter addresses the second research question:

RQ 2: How does an altering community focus influence the contribution activity of specialized members?

The three member groups (Commercials, Micro-entrepreneurs and Privates) focus only 30 to 40 %[47] of their contributions on their specialized categories. Nevertheless, the frequency of threads from these categories should influence their relative activity significantly as predicted in the hypotheses chapter. This assumes that an increased frequency of threads specialized in is followed by an increased relative activity of the respective member group and accordingly an increased frequency of the contrary threads is followed by a decreased relative activity led to eight operationalized hypotheses.

5.4.1 Descriptive analysis

The change in the relative importance ($\Delta p_{j,t}$) of the three categories Operations, Sales and Marketing as well as the Member group activity changes ($GAC_{k,t}$) for Commercials, Micro-entrepreneurs and Privates are shown in the following table:

---

[47] Commercials with 31.3 % on Operations and Sales; Micro-entrepreneurs with 41.7 % on Marketing and Sales; Privates with 32.5 % on Marketing

Category/ Member group	N	Min	Max	Mean	Standard Deviation
Operations	61	-0.38	0.28	0.0003	0.12
Sales	61	-0.23	0.37	-0.0019	0.12
Marketing	61	-0.55	0.71	-0.0037	0.21
Commercials	61	-1.00	1.00	0.0066	0.51
Micro-entrepreneurs	61	-0.80	0.67	-0.0445	0.32
Privates	61	-1.00	1.00	0.0076	0.43

Table 17: $\Delta p_{j,t}$ of categories and $GAC_{k,t}$ of community member groups

Even though the complete set of data features 62 months, the sample size of $\Delta p_{j,t}$ and $GAC_{k,t}$ is limited to 61. Both variables present a relative change in a respective month compared to the previous month and therefore, the second month represents the first possible value. As outlined explicitly when deriving the variables, the value of both variables can range from -1 to +1. The importance of the Operations and Sales category changes moderately considering a one-month period. The maximum change is 0.38 respectively 0.37 and the relatively low standard deviation of both variables indicates a fairly constant relative importance of both categories. In contrast, the percentage share of Marketing threads differs substantially considering a month-to-month change, ranging from -0.55 to +0.71 resulting in an almost doubled standard deviation compared to the other categories. Regarding the relative activity changes of the different community member groups, the Commercials and Privates show a similar behavior. Their maximum and minimum group activity change ($GAC_{k,t}$) value indicates that at least in one month all members of the respective group increased or decreased their relative activity simultaneously. On thecontrary, the Micro-entrepreneurs seem to contribute to the community more continuously pointed out by the lower standard deviation and the lacking of -1 or +1 as maximum, respectively minimum value.

5.4.2   Data distribution

The distribution characteristic of $\Delta p_{j,t}$ of categories and $GAC_{k,t}$ of community member groups, determines the kind of statistical assessment to be conducted. The following figure shows the histograms of all variables applied to support the hypotheses from this chapter:

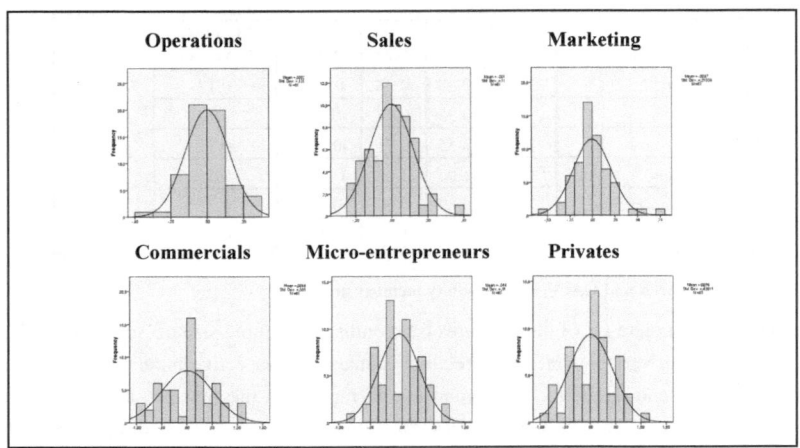

Table 18: Histograms $\Delta p_{j,t}$ of categories and $GAC_{k,t}$ of community member groups

Based on the graphical assessment all six variables seem to be approximately normally distributed. Considering conducting linear regressions in this part of the empirical study, which compulsorily require normally distributed data, further assurance regarding the distribution is required. Therefore, the skewness and kurtosis of all variables are evaluated via SPSS and shown in the following table:

Category/	Skewness		Kurtosis	
Member group	Statistic	Std. Error	Statistic	Std. Error
Operations	-0.298	0.306	0.840	0.604
Sales	0.284	0.306	0.411	0.604
Marketing	0.629	0.306	2.259	0.604
Commercials	-0.079	0.306	-0.595	0.604
Micro-entrepreneurs	0.061	0.306	-0.428	0.604
Privates	-0.075	0.306	-0.389	0.604

Table 19: Skewness and kurtosis of $\Delta p_{j,t}$ of categories and $GAC_{k,t}$ of groups

Following Miles et al. (2010) a set of data can be approximated as normally distributed if the value for skewness does not exceed one or minus one. Thus, all variables can be treated as normally distributed. Nevertheless, the Kolmogorov-Smirnov test on normal distribution is additionally conducted and the results summarized in the following table:

Category/ Member group	N	Kolmogorov-Smirnov Z	Significance level
Operations	61	0.597	0.868
Sales	61	0.709	0.696
Marketing	61	0.774	0.588
Commercials	61	0.948	0.330
Micro-entrepreneurs	61	0.579	0.891
Privates	61	0.747	0.632

Table 20: Results from K-S test for $\Delta p_{j,t}$ of categories and $GAC_{k,t}$ of groups

The significance levels for all variables exceed 0.05 and therefore the data can finally be treated as normally distributed and linear regression models can be computed.

5.4.3 Linear regression models

To measure the influence of changes in the category importance on the relative activity of certain member groups, linear regression models are computed. Considering the data and the derived hypotheses, a simple model with only one independent and one dependent variable is chosen:

$$Y = b_0 + b_1 X + u$$

where Y denotes the dependent variable ('$GAC_{k,t}$'), b_0 and b_1 the unknown regression coefficients, X the independent variable ('$\Delta p_{j,t}$') and u the residuals. As mentioned in the fourth chapter, the threads to the mailing list are posted from the Central Organizer or some Coordinators. Hence, the relative importance of the categories is independent of the contributions of the analyzed three member groups, Commercials, Micro-entrepreneurs and Privates. Before computing eight different regression lines[48], a check on correlations between the dependent and independent variables seems to be expedient. Non-correlated pairs are excluded from the regression analyses to reduce the computation effort:

---

[48] One for each of the eight hypotheses

			Δp		
			Operations	**Sales**	**Marketing**
GAC	**Commercials**	Pearson Correlation	-0.023	0.079	0.053
		Sig. (2-tailed)	0.86	0.543	0.686
		N	61	61	61
	Micro-entrepreneurs	Pearson Correlation	0.041	**0.329**	0.113
		Sig. (2-tailed)	0.753	**0.01**	0.388
		N	61	**61**	61
	Privates	Pearson Correlation	**-0.331**		0.082
		Sig. (2-tailed)	**0.009**		0.528
		N	**61**		61

Table 21: Pearson coefficients $\Delta p_{j,t}$ of categories and $GAC_{k,t}$ of groups

Referring to the results for the Pearson correlation coefficients, only two of the eight predicted coherences show a sufficient significance level[49]. Community members belonging to the group of Commercials seem to entirely ignore the importance of certain value chain categories. They do not adjust their relative activity in accordance with relative focus changes of the community discussion. The Pearson correlation coefficients for this group barely vary from zero and therefore negate the coherence between the relative importance of categories and their relative contribution activity. The relative contribution activity of Micro-entrepreneurs is positively driven by the importance of the Sales category. Nevertheless, neither influences Operations their relative activity negatively nor their other specialization category Marketing the relative activity positively, as presumed. On the contrary, for the community members from the Privates group no positive correlation between their specialization area Marketing and their relative activity level can be identified. Their relative activity level is only negatively correlated with the importance of the Operations category as predicted. Summarizing, six of the eight derived hypotheses for this part of the empirical study need to be rejected even before starting with the linear regression model.

H 2.1: Commercials increase their relative contribution activity with an increased importance of Operations threads

H 2.2: Commercials increase their relative contribution activity with an increased importance of Sales threads

---

[49] Both on a very strong 0.01 significance level

H 2.3: Commercials decrease their relative contribution activity with an increased importance of Marketing threads

H 2.4: Micro-entrepreneurs increase their relative contribution activity with an increased importance of Marketing threads

H 2.6: Micro-entrepreneurs decrease their relative contribution activity with an increased importance of Operations threads

H 2.7: Privates increase their relative contribution activity with an increased importance of Marketing threads

For the regression model only the combination of '$GAC_{Micro-entrepreneurs}$' and '$\Delta p_{Sales}$' as well as '$GAC_{Privates}$' and '$\Delta p_{Operations}$' are considered. To obtain sufficient confidence in the model results, various assumptions need to be validated. Considering the characteristics of the chosen linear regression model and the data, four assumptions needs to be covered (Harrell 2002; Backhaus 2008): (1) The dependent and independent variable are linearly interrelated; (2) the residuals show a constant variance (homoscedasticity); (3) the residuals are normally distributed; (4) no autocorrelation exists. The fourth assumption takes on paramount importance especially in time series data analysis. Therefore, this assumption is validated before discussing the other three assumptions to hopefully reduce the required effort.

To finally support the fifth hypothesis of this chapter, the regression model is evaluated with '$GAC_{Micro-entrepreneurs}$' as dependent and '$\Delta p_{Sales}$' as independent variable. The following table shows the model summary and the results from the Durbin-Watson test on autocorrelation:

Model Summary[b]

Model	R	R Square	Adjusted R Square	Std. Error of the Estimate	Durbin-Watson
1	,329[a]	,109	,093	,30342	2,086

a. Predictors: (Constant), Sales
b. Dependent Variable: Micro

Coefficients[a]

Model		Unstandardized Coefficients		Standardized Coefficients	t	Sig.
		B	Std. Error	Beta		
1	(Constant)	-,043	,039		-1,101	,275
	Sales	,879	,328	,329	2,681	,010

a. Dependent Variable: Micro

Table 22: Model summary regression model for 'Δp_{Sales}' and '$GAC_{Micro-entrepreneurs}$'

The result of the Durbin-Watson test is represented by a coefficient d, which is the aggregation of the differences between the residuals of succeeding observation points. A value of two indicates no autocorrelation at all, a value of zero a complete positive autocorrelation and a value of four full negative autocorrelation. To successfully support H_0: 'The observation points are not auto correlated' the allowed range for the value of d is between d_U and $(4-d_U)$ (Backhaus 2008). For a sample size of n=61 and a 95% confidence level $d_U=1.62^{50}$ is required. The determined value of 2.09 is below the upper barrier of 2,38 and therefore autocorrelation can be precluded. The model is significant on a 0.01-level ($\alpha = 0.010$) with an R square of 11%. Initially this appears rather small and it is obvious that other factors beyond the relative importance of the Sales category affect the activity level of the Micro-entrepreneurs. Nevertheless, considering the high share of mails sent to other categories (more than 80%), the importance provides a recognizable impact on the relative activity of the Micro-entrepreneurs. After determining a sufficient significance level and precluding autocorrelation, the other three assumptions need to be validated. Therefore, the next figure shows a plot of the dependent vs. independent variable (1), a plot of the residuals vs. predicted values (2), and a check for normal distribution of the residuals (3):

---

[50] Cf. Backhaus 2008, p. 567

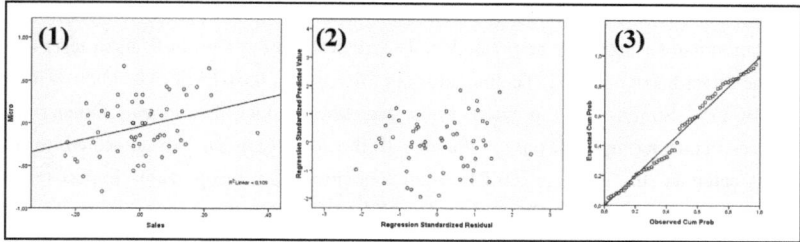

Figure 19: Validation assumptions regression model, Sales and Micro-entrepreneurs

The first scatter plot including the linear regression line proves that a linear relationship between the dependent and independent variable can be assumed. The second plot proves the existence of homoscedasticity for the residuals by indicating constant variances. Lastly, the deviations from the straight line are from minor characteristics and therefore the residuals follow a normal distribution. After validating all assumptions, the fifth hypothesis is finally supported:

H 2.5: Micro-entrepreneurs increase their relative contribution activity with an increased importance of Sales threads

To verify the last hypothesis of this chapter, which has not already been rejected after observing the Pearson correlation coefficients, the model is evaluated with '$GAC_{Privates}$' as the dependent and '$\Delta p_{Operations}$' as the independent variable. The following table shows the model summary and the results from the Durbin-Watson test on autocorrelation:

Model Summary[b]

Model	R	R Square	Adjusted R Square	Std. Error of the Estimate	Durbin-Watson
1	,331[a]	,110	,095	,40927	1,916

a. Predictors: (Constant), Operations
b. Dependent Variable: Privates

Coefficients[a]

Model		Unstandardized Coefficients		Standardized Coefficients	t	Sig.
		B	Std. Error	Beta		
1	(Constant)	,008	,052	.	,153	,879
	Operations	-1,184	,439	-,331	-2,696	,009

a. Dependent Variable: Privates

Table 23: Model summary regression model for '$\Delta p_{Operations}$' and '$GAC_{Privates}$'

99

Considering the allowed range for the value of d, which has been previously determined, autocorrelation can successfully be precluded. The value from the Durbin-Watson test (1.92) exceeds the lower barrier of 1.62. The model is significant on a 0.01-level ($\alpha = 0.009$) with an R square of 11%. Similar to the previous linear regression model, this appears rather small. Nevertheless, considering that only 12% of all threads discussed are allocated to the Operations category, the importance of this category provides a recognizable impact on the relative activity of the Privates. Once again to validate the other three assumptions the next figure shows a plot of the dependent vs. independent variable (1), a plot of the residuals vs. predicted values (2), and a check for normal distribution of the residuals (3):

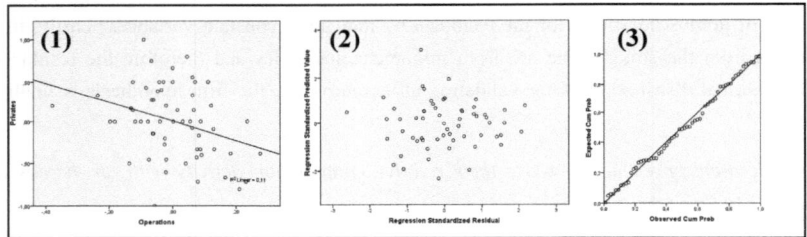

Figure 20: Validation assumptions regression model, Operations and Privates

The three plots support the assumptions and finally the last hypothesis of this chapter is supported.

H 2.8: Privates decrease their relative contribution activity with an increased importance of Operations threads

5.4.4 Variation of model

Two hypotheses are supported with a strong significance level of 0.01. Nevertheless, six hypotheses need to be rejected. Considering this limited support of the predicted behavior, slight variations of the standard model might unveil new additional insights concerning the contribution activity of specialized community members. Referring to empirical findings of other researchers, the contribution activity is strongly influenced by the community group size and the communication volume. In larger systems, the hope of reciprocity is dramatically diluted simply by the greater distance between the contributor and the beneficiary (Thorn and Connolly 1987). This leads to lower participation and finally community leaving of certain members. Schoberth et al. observe the same behavior caused by a state that is called 'information overload', where the communication volume is too high to be successfully
100

absorbed by the community members (Schoberth et al. 2003). Returning to the case of 'Premium', an individual community member might start to consider changing his contribution activity if the monthly communication volume exceeds a specific barrier, a tipping point. With a higher number of contributions, the member is forced to invest more time to drive the community communication significantly (von Hippel and von Krogh 2003). In that case, the importance of their liked or disliked categories could be of greater relevance to them than in months with lower communication volume. To examine this effect, the sample size is varied based on the number of contributions made within one month. The following table shows the Pearson correlation coefficients if considering only the 50[51] (~83% of all months), 41 (~67%) and 31 (~50%) months with the highest communication volume:

---

[51] Normally 51 months represent 5/6, respectively 83% of all months, but in the data set three months have the same number of mails and therefore the sample size is slightly varied.

			Δp		
			Operations	Sales	Marketing
GAC	Commercials$_{83,3}$	Pearson Correlation	0,000	0.135	-0.029
		Sig. (2-tailed)	0.998	0.349	0.841
		N	50	50	50
	Micro-entrepreneurs$_{83,3}$	Pearson Correlation	-0.025	**0.257**	0.093
		Sig. (2-tailed)	0.863	**0.071**	0.52
		N	50	**50**	50
	Privates$_{83,3}$	Pearson Correlation	**-0.399**		0.073
		Sig. (2-tailed)	**0.004**		0.613
		N	**50**		50
GAC	Commercials$_{66,7}$	Pearson Correlation	-0.068	0.166	0.028
		Sig. (2-tailed)	0.671	0.3	0.862
		N	41	41	41
	Micro-entrepreneurs$_{66,7}$	Pearson Correlation	-0.215	0.238	0.208
		Sig. (2-tailed)	0.178	0.133	0.193
		N	41	41	41
	Privates$_{66,7}$	Pearson Correlation	**-0.479**		0.2
		Sig. (2-tailed)	**0.002**		0.193
		N	**41**		41
GAC	Commercials$_{50}$	Pearson Correlation	-0.129	0.294	-0.18
		Sig. (2-tailed)	0.491	0.108	0.334
		N	31	31	31
	Micro-entrepreneurs$_{50}$	Pearson Correlation	-0.206	0.196	**0.315**
		Sig. (2-tailed)	0.266	0.291	**0.085**
		N	31	31	**31**
	Privates$_{50}$	Pearson Correlation	**-0.604**		0.199
		Sig. (2-tailed)	**0,000**		0.284
		N	**31**		31

Table 24: Information overload analysis

Even with this variation of the normal model the group of Commercials do not show any contribution activity change with a sufficient significance level. Nevertheless, the value of the correlation coefficients is higher if only months with a high communication volume are considered. The previously identified significant cohesion between the importance of the Sales category and the relative activity change of the Micro-entrepreneurs disappears with an increased communication volume. The positive correlation between changes in their activity and the Marketing category for the months with the highest communication volume is only supported with a rather low significance level of 0.1. For the last group, the Privates, no additional significant correlation can be identified. However, the negative effect of Operation threads on their relative contribution activity is tremendously amplified with an increasing communication volume. Considering only months with the highest activity, the correlation coefficient reaches a very strong value of more than -0.6.

An additional variation is to consider a lagged impact of the independent on the dependent variable. This well-established approach in time series analysis[52] regards the possible time lag in each individual's behavior. The community member might need some time to recognize the changes in the importance of certain categories and react accordingly to them. Therefore, in the following a time lag of one month for the group activity change variables (GAC_{t+1}.) is incorporated into the analysis. Necessarily the sample size is reduced to 60. The following table shows the Pearson correlation coefficient for Δp and the time lagged GAC:

			Δp		
			Operations	Sales	Marketing
GAC	Commercials$_{t+1}$	Pearson Correlation	0.07	0.011	0.058
		Sig. (2-tailed)	0.595	0.935	0.657
		N	60	60	60
	Micro-entrepreneurs$_{t+1}$	Pearson Correlation	-0.008	-0.143	-0.188
		Sig. (2-tailed)	0.953	0.277	0.206
		N	60	60	60
	Privates$_{t+1}$	Pearson Correlation	-0.068		0.055
		Sig. (2-tailed)	0.606		0.678
		N	60		60

Table 25: Time-lag analysis

All eight combinations between GAC and Δp show no significant correlations. The previously identified significant correlations between Micro-entrepreneurs and Sales and Privates and Operations are not verified in this data set. Therefore, a lagged impact on the contribution activity as presumed is not verified.

Referring to the results from the linear regressions and the two variations of the model the three member groups act differently. The commercial members completely ignore the importance of certain categories and do not align their activity based on an altering community focus. The group of Micro-entrepreneurs seems to be driven by capturing additional benefits as their contribution activity is positively influenced by the importance of their specialization category. In contrast, the private community members try to avoid high participation costs as their activity is negatively driven by their opposed categories. This effect is further amplified by a high monthly communication volume.

---

[52] E.g. the impact of patent applications on corporate performance (Ernst 2001)

5.5 Transfer of special knowledge into the community

The last chapter of the results from the quantitative analyses addresses the third research question:

Research question 3: How is knowledge from specialized individuals integrated permanently into the community?

For the first time the fourth group *Coordinators* is included into the analyses. As predicted in the third chapter, this group accomplishes the task to integrate knowledge into the community permanently. The group act like Men-on-the-inside (MOI) and hold strong communication ties to the specialized groups Commercials, Micro-entrepreneurs and Privates. As previously explained, a high CoCit-Score between the different individual community members is equivalent to a tight communication. It is predicted that the CoCit-Scores between each of the specialized member groups to the Coordinators exceed the CoCit-Scores to all other groups.

5.5.1 Descriptive analysis

The following table shows the mean CoCit-Scores between all four member groups excluding the central organizer:

Mean CoCit-Score (10^-3)	Coordinators	Commercials	Micro-entrepreneurs	Privates
Coordinators	18.08	11.80	11.55	10.73
Commercials		8.84	8.51	6.16
Micro-entrepreneurs			10.63	6.73
Privates				6.81

Table 26: Mean CoCit-Scores between member groups

Each of the groups shows the highest mean CoCit-score to the Coordinators as previously predicted. The Commercials, Micro-entrepreneurs and Privates hold the second highest score to their respective member group. This finding was also expected, as one impulse for specialization is the focus on threads fostering the collaboration with peer-group minded members. The CoCit-Scores also reveal additional insights, such as the relatively low score level between the Privates and all other member groups. This group seems to contribute to fewer threads or multiply to the same thread both resulting in fewer co-citations and lower

CoCit-Scores. On the other hand the CoCit-Scores between the Micro-entrepreneurs and the other member groups are quite high, especially to their own member group (10.63×10^{-3}) almost exceeding the score to the Coordinators (11.55×10^{-3}). However, a statistical assessment is required to confirm a sufficient significance level of the observed differences.

5.5.2 Data distribution

To verify the three hypotheses, the CoCit-Scores of the three member groups Commercials, Micro-entrepreneurs and Privates to the other four member groups must be included in the statistical assessment. Therefore, ten distributions need to be characterized to identify the appropriated statistical test and are shown in the next histograms:

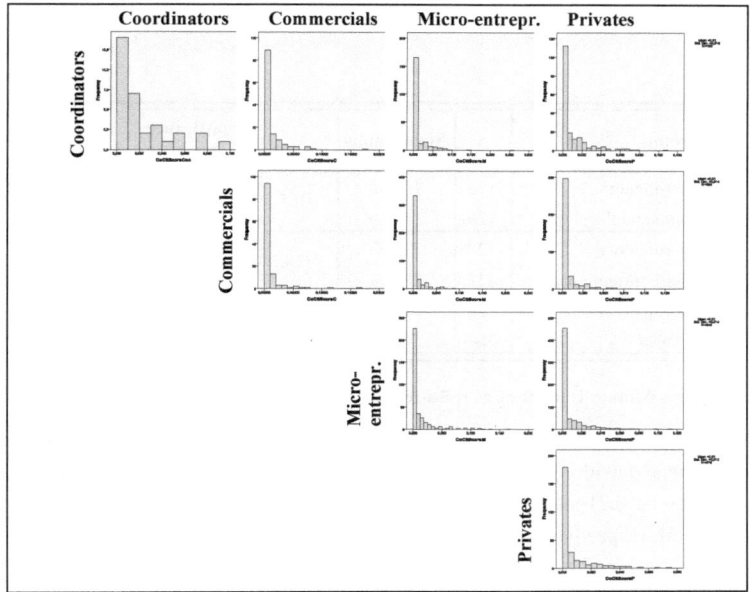

Figure 21: Histograms CoCit-Scores

Normality is rejected for all CoCit-Scores between the different member groups. All distributions show an extreme right skewness caused by the high amount of zero and small values. As discussed previously these values represent a low level of communication between the individuals and therefore contribute to a comprehensive understanding of the community dynamics. Due to the non-normality, only non-parametric tests are applied. Similar to the

105

result chapter for the contribution focus, central tendencies between two samples are compared and according to the independencies of the samples, the Mann-Whitney U test is appropriate.

5.5.3 Mann-Whitney-U test

To verify that the three groups hold the strongest communication ties to the group of Coordinators, three Mann-Whitney U tests are required for each group. It has to be proven that the CoCit-Scores between group A and the Coordinators exceed the scores between group A and the groups A, B and C.

The three following tables (table 27 to table 29) show the results for the Commercials, Micro-entrepreneurs and Privates:

	Groups	N	Mean ranks	U	Significance level
CoCit-Scores Commercials to...	Coordinators	128	133.98	6467	0.020
	Commercials	120	114.39		
	Coordinators	128	318.83	22742	0.001
	Micro-entrepreneurs	432	269.14		
	Coordinators	128	295.73	19555	0.000
	Privates	384	243.42		

Table 27: Results Mann-Whitney U test for CoCit-Scores Commercials to...

The CoCit-Scores between the *Commercials* and the *Coordinators* exceed the scores to all other groups. Compared with their peer group a 0.05-significance level is achieved ($\alpha = 0.020$), surpassed by an 0.01-significance level achieved for the comparison with the other two groups ($\alpha = 0.001$, respectively $\alpha = 0.000$). According to these results the hypothesis:

H 3.1: Commercials show higher CoCit-Scores to Coordinators than to any other community member group

is fully supported.

	Groups	N	Mean ranks	U	Significance level
CoCit-Scores Micro-entrepreneurs to...	Coordinators	216	343,00	42659	0.045
	Commercials	432	315.25		
	Coordinators	216	280.80	37217.5	0.692
	Micro-entrepreneurs	351	285.97		
	Coordinators	216	454.52	65228.5	0.094
	Privates	648	425.16		

Table 28: Results Mann-Whitney U test for CoCit-Scores Micro-entrepreneurs to...

The Micro-entrepreneurs show a contrary communication pattern. They do show an intensified communication with the Coordinators compared to the Commercials and Privates but not compared to their own peer group. Additionally, the findings are supported on a rather low significance level of 0.05 and 0.1 ($\alpha = 0.045$ compared with Commercials and $\alpha = 0.094$ with Privates) and the Mann-Whitney U test even indicates a higher mean rank for their peer group (285.97) in comparison with the Coordinators (280.80). Based on these results the hypothesis:

H 3.2: Micro-entrepreneurs show higher CoCit-Scores to Coordinators than to any   other community member group

needs to be rejected.

	Groups	N	Mean ranks	U	Significance level
CoCit-Scores Privates to...	Coordinators	192	331.82	28547	0.000
	Commercials	384	266.84		
	Coordinators	192	477.90	51188	0.000
	Micro-entrepreneurs	648	403.49		
	Coordinators	192	256.59	22254	0.002
	Privates	276	219.14		

Table 29: Results Mann-Whitney U test for CoCit-Scores Privates to...

The findings for the third group Privates are in accordance with the results for the Commercials. The Privates show higher scores to the Coordinators than to any other group on a very strong 0.01-significance level ($\alpha = 0.000$, $\alpha = 0.000$ and $\alpha = 0.002$). Therefore the third hypothesis of this chapter is fully supported:

H 3.3: Privates show higher CoCit-Scores to Coordinator than to any other   community member group

These results prove the importance of members with dedicated roles for the knowledge transfer. Even though commercial and private members communicate intensively with their own peers, their strongest communication takes place with the Coordinators. Contrary, the Micro-entrepreneurs communicate most intensively with their own peers.

5.6 Discussion and implications for further proceeding

The first and main part of the empirical study revealed unpredicted behavior and additional findings for the member specialization phenomenon. The phenomenon seems to be multi-layered, more diverse than expected and further investigations are required. However, the results from the quantitative analyses are summarized before discussing these findings and illustrating the further proceeding to capture additional aspects of member specialization.

5.6.1 Summary of results from the quantitative part of the empirical study

The investigation of differences in the contribution focus of the community member groups delivered mixed results. For the 'Overall sample' containing all 103 members, two of six hypotheses could be supported on a sufficient significance level. The Micro-entrepreneurs and the Privates show both a significantly higher focus on the Marketing category than the Commercials. The results for the 'Mature sample' containing only the members exceeding 10% of the mean activity are of superior quality, meaning they show improved significance levels. Four of six hypotheses are supported on a sufficient significance level ($\alpha = 0.1$ or better). Especially the comparison of the Commercials and Privates unveiled a clear contribution focus of both groups by supporting all three hypotheses. Whereas the Commercials specialize on the Operations and Sales categories, the Privates contribute mostly to Marketing. In contrast, the Micro-entrepreneurs show a less intense contribution focus on certain categories, indicated by verifying only one out of three hypotheses including this community member group[53].

The second cluster of results, the influence of changes in the importance of certain categories on the relative activity level of different member groups, is by far the most dissatisfying. From the derived eight hypotheses, only two could be finally supported by the conducted analyses. The Commercials seem to align their activities in no sense to changes of the community focus. The activity of the Micro-entrepreneurs is not influenced by the importance of the Operations or Marketing category as previously predicted but the importance of their favored Sales category positively influences their contribution activity. On the contrary, the Privates' contribution activity is not positively influenced by their favored Marketing category. However, the predicted negative impact of a relative increase in Operations threads is fully supported and strongly amplified with an increase in communication volume. In contrast to other time series analyses a time lagged impact of the independent on the

---

[53] Focus on Marketing compared with the Commercials

dependent variable could not be identified. Changes in the importance of certain categories in one month have absolutely no impact on the relative contribution activity of the different member groups in the following month.

By observing the communication and collaboration between the four different member groups[54], the special role of the Coordinators group can be verified, even though only two of the three hypotheses are supported. The strong communication ties between the Coordinators and the Commercials respectively the Privates are supported on a very strong significance level[55]. The Micro-entrepreneurs pursue a more intense communication to the Coordinators compared to the Commercials and Privates but the most intensive knowledge exchange takes place within their own peer-group.

5.6.2 Discussion and further proceeding

The first part of the empirical study shows mixed results and additional findings not predicted before in the phenomenological and theoretical part of this thesis. Especially the group of Micro-entrepreneurs refuses some of the predicted behavioral patterns. Considering all quantitative results from the empirical study, the different behavioral patterns of the various community member groups have been underestimated. Their different individual familiarities with the community outcome seem to generate more insights for the specialization phenomenon than just a different contribution focus.

Concerning the contribution focus, Commercials and Privates concentrate their efforts as previously predicted. On the other hand, the Micro-entrepreneurs show a contribution focus on the Marketing category compared to the Commercials, but also on Product Development and Operations compared to the Privates. Particularly the major differences to the Privates are astonishing, since these two groups are rather similar only distinguished by the additional entrepreneurial tasks of the Micro-entrepreneurs. Even the group sizes are comparable (n=36 vs. n=31) and their mean activity level and its standard deviation are almost equal (42 mails and standard deviation of 81 vs. 42 mails and standard deviation of 80). Nevertheless, the Micro-entrepreneurs are more focused on product-related categories (value-chain activities) than the Privates and even the Commercials. The superior results for the 'Mature sample' indicate that the contribution focus seems to intensify over time. Von Krogh et al. have already questioned this dynamic aspect of specialization (von Krogh et al. 2003).

---

[54] The last research question concerning the integration of special knowledge into the community additionally addresses the fourth group Coordinators.
[55] Mostly within the range of $\alpha = 0.002$ and $\alpha = 0.000$

Referring to the results for the contribution activity all three community member groups show a different behavior when the same time periods are analyzed. The Commercials completely ignore the importance of certain categories, even the ones specialized in, when considering aligning their relative contribution activity level. However, with an increased communication volume the tendency to adjust their activity to the importance of certain categories rises but a sufficient significance level for a reasonable support is not achieved. Similar to the Commercials, the contribution activity of the Micro-entrepreneurs is not negatively influenced by the frequency of threads from certain opposing categories. Nevertheless, their activity is positively driven by their specialization category Sales although this effect disappears with a higher monthly communication volume. In contrast, the activity of the Privates is only negatively driven by categories opposed to their specialization area. This effect is further amplified as the communication volume increases. However, all three member groups have in common that they do not adjust their contribution activity with a time lag. This indicates rather an immediate behavioral change to altered circumstances than following a long-term strategy.

The results from the co-citation analysis prove the important role of the member group Coordinators. They integrate knowledge into the community on a permanent basis by helping to transform tacit into explicit knowledge by holding strong communication ties to the majority of the Commercials and Privates. Contrary to the Micro-entrepreneurs, they show their strongest communication ties to their own peer-group and therefore they are less 'controllable' by the Coordinators. They seem to occupy an encapsulated position within the community. However, the observed CoCit-Scores provide another interesting finding. The Privates show lower scores to all other groups, which could be a sign of a higher specialization degree. In contrast to the other groups, they seem to contribute repeatedly to the same threads resulting in a more unbalanced participation.

Summarizing four different tasks need to be part of further analyses:

- Reveal differences in the specialization degree.

- Show the development of the contribution focus over time.

- Review the encapsulated position of the Micro-entrepreneurs.

- Unveil first indications for the rationales behind the unpredicted behavioral patterns, especially for the differences in the adjustment of the contribution activity following an alteration of the community discussion focus.

6 Qualitative analyses and results

The quantitative study provided interesting additional research opportunities to enrich the research contribution to member specialization in OIC. The goal of this second part of the empirical study is to detail some of the quantitative results, initially support the new findings and identify the rationales behind the unpredicted behavior. For this purpose, only qualitative analyses are applied in this part of the empirical study. Although most of the mixed-method studies initially start with a qualitative inquiry, the chosen sequence in this study seems to be appropriate. "Qualitative inquiry might also follow quantitative analysis. Such an activity might appear methodologically incongruous to some, but is particularly useful when a researcher wishes to (a) attempt to explain the existence of an unexpected pattern in the data, or (b) attempt to uncover the mechanism(s) that create that unexpected pattern" (Shah and Corley 2006, p. 1832). The different analyses are based on the finalized data set from the fourth chapter. The data is displayed in various systematic formats, which allows the researchers to draw conclusions and support the newly identified findings (Miles and Huberman 1994). Different qualitative methods are applied including several semi-structured interviews with different individuals from the three member groups Commercials, Micro-entrepreneurs and Privates. The results from this chapter should support and help to explain the new aspects identified for member specialization in OIC and especially the extraordinary role of the Micro-entrepreneurs as mentioned in the fifth chapter. Two different coefficients for determining the degree of specialization are applied. The change in the contribution foci over time revealed during the first part of the study is treated by investigating the joining behavior of new community members. To analyze the special position of the Micro-entrepreneurs within the community, a graphical assessment is conducted after sketching the social network spanned by all members. Therefore, the social network analysis (SNA) is described in detail since the application of SNA is rather complex and different key measures need to be explained to fully understand and interpret the outcomes. Finally, the results from the interviews are specified to show additional insights concerning the drivers of each member group and to unveil some rationales behind their unpredicted behavior.

6.1 Homogeneity analysis

The evidence taken from the co-citation analysis, that the specialization degree also depends on the affiliation to a specific community member group, is assessed by conducting two different homogeneity analyses: The mails per thread ratio and the Gini coefficient.

6.1.1 Mails per thread ratio

Community members showing a high degree of specialization contribute rather to a few than simultaneously to many modules (von Krogh et al. 2003). Adopting this definition, the mails per thread ratio functions as an indicator for the degree of specialization. Each individual community member decides to disperse his total number of contributions to a certain number of threads. Contributing repeatedly to the same thread means a high degree of specialization whereas spreading the contributions to more threads indicates a low degree of specialization and vice versa a high degree of generalization. Based on the different mean CoCit-Scores of the three member groups, a difference in the specialization degree has been assumed. The following table shows the mails per thread ratio by the respective groups:

Member group	N	Min	Max	Mean	Standard Deviation
Commercials	26	1.00	1.84	1.19	0.24
Micro-entrepreneurs	36	1.00	1.82	1.24	0.23
Privates	31	1.00	3.00	1.45	0.50

Table 30: Mails per thread ratio for member groups

Privates show a higher degree of specialization based on their higher mails per thread ratio. As previously predicted, they concentrate their contributions on fewer threads. This result could indicate a higher social motivation of the Privates. Content driven members are pleased to contribute their solution, knowledge or opinion to selected threads. On the contrary, socially driven members are motivated by interacting with others leading to intense discussions about the same thread.

6.1.2 Gini coefficient

The degree of specialization can additionally be determined by measuring the contribution dispersion of each community member on the different thread categories. Similar to the previously outlined definition, highly specialized members concentrate most of their contributions to a relatively low share of categories. On the contrary, members with a high generalization degree spread their contributions more balanced to a relatively high share of categories. The Gini-coefficient represents an appropriate and well-established measurement of this inequality. Developed by Corrado Gini in 1912, this coefficient is primarily applied for measuring income inequalities in specific countries using the Lorenz curve. "The Lorenz curve relates the cumulative proportion of income units to the cumulative proportion of income received when units are arranged in ascending order of their income" (Kakwani 1977, p. 719). For this application, the Gini index "is the ratio of the area between the Lorenz curve and the 45° line to the area under the 45° line" (Gastwirth 1972, p. 307) as shown on the left side of the following figure:

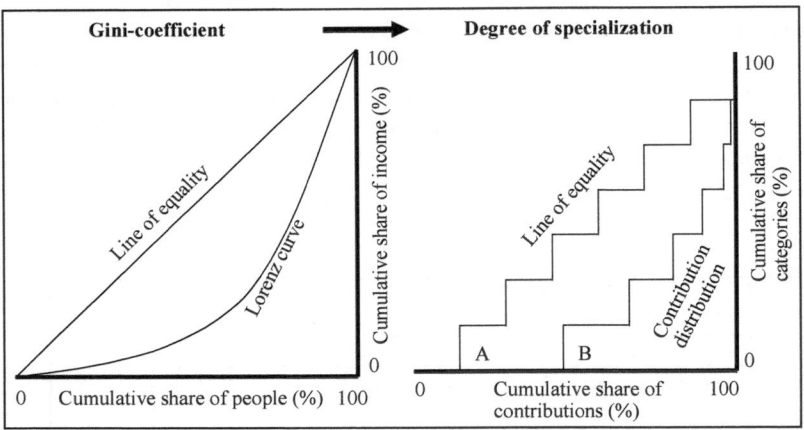

Figure 22: Gini coefficient and adaptation for measuring specialization degree

The 45° line represents the line of equality, meaning the shape of the Lorenz curve if the income is completely equally distributed among all individuals. The area between this line and the Lorenz curve is called the area of concentration (Gastwirth 1972). The inequality rises with an increase of the area of concentration. The Gini coefficient G

$$G = 1 - 2\int_0^1 L(x)$$

takes values from zero to one where an index value of one represents a total inequality as the area under the Lorenz curve is zero, vice versa a value of zero indicates total equality. Even though the coefficient is commonly used to analyze income distributions, its application fields are diverse, e.g. to measure the geographical concentration of production and innovation (Audretsch and Feldmann 1996) or even the diversity of bacterial soil communities (Harch et

114

al. 1997). Furthermore, it is not limited to continuous functions and can be applied for discrete probability functions. Therefore, the Gini coefficient can be used for measuring the degree of specialization as shown on the right side of Figure 22: Gini coefficient and adaptation for measuring specialization degree. Total equality occurs if an individual community member disperses the contributions equally to each of the categories. This results in a contribution share of $1/n$ for each of the categories with n = number of categories and is illustrated by the line of equality. The contribution distribution graph represents the actual cumulative share of contributions to the different categories for each individual. The higher the specialization degree, the higher the contribution shares to a smaller number of categories, the smaller the area B under the contribution distribution graph. This again necessarily leads to a higher Gini coefficient:

$$G = \frac{A}{A+B} \quad \rightarrow \quad G = 1 - \frac{B}{A+B}$$

For a discrete probability function with $p_1, p_2, ..., p_n$ as probabilities for each category arranged in ascending order and n categories the Gini coefficient is represented by,

$$G = 1 - \frac{2}{(n-1)}\left(\sum_{i=1}^{n-1} p_i + \sum_{i=1}^{n-2} p_i + ... + p_1\right)$$

due to $A + B = \frac{1}{n}\left(\frac{(n-1)}{n} + \frac{(n-2)}{n} + ... + \frac{1}{n}\right) = \frac{1}{n^2}\left(\frac{n(n-1)}{2}\right) = \frac{(n-1)}{2n}$

and $B = \frac{1}{n}\left(\sum_{i=1}^{n-1} p_i + \sum_{i=1}^{n-2} p_i + ... + p_1\right)$

Based on the contribution distribution consisting of the seven category probabilities, the Gini coefficient for each individual community member is calculated and summarized on group level shown in the following table:

Member group	N	Min	Max	Mean	Standard Deviation
Commercials	26	0.24	1.00	0.59	0.25
Micro-entrepreneurs	36	0.19	1.00	0.58	0.22
Privates	31	0.22	1.00	0.64	0.23

Table 31: Mean Gini coefficient of each member group

The Privates show the highest mean Gini coefficient (G=0.641) and therefore the highest degree of specialization. The difference between the other two groups (G=0.594 for the Commercials and G=0.576 for the Micro-entrepreneurs) is rather small. These findings are in accordance with the mails per thread ratio, which also showed the highest specialization degree for the Privates. This group seems to not only concentrate their contributions on fewer threads but also on fewer categories. Nevertheless, the maximum values of the Gini coefficient additionally display that at least one individual member of each group is extremely specialized. These members concentrate all their contributions completely on threads from one category.

6.2 Initial contribution analysis

The initial contribution made by new community members is commonly used as an analyzing unit for investigating the joining process of members. The analyses of other researchers deal with the content of this contribution (von Krogh et al. 2003; Herraiz et al. 2006) or the recipients of it (Qureshi and Fang 2010). To show the misled specialization of new community members, the content of their initial contribution is the adequate measurement. Empirical findings in the quantitative part of this study showed a stronger contribution focus of the 'Mature sample'. The members of that sample focus their contributions mostly as predicted. Their initial mails can indicate whether this specialization was in place right from the very beginning of their community tenure or not. For this analysis only members from the 'Mature sample' are included who additionally made their first contribution later than four month after the start of the observation period in March 2005. This ensures that only members who newly joined are involved. Figure 23: Initial contributions of 'Mature sample' shows the results from the initial contribution analysis:

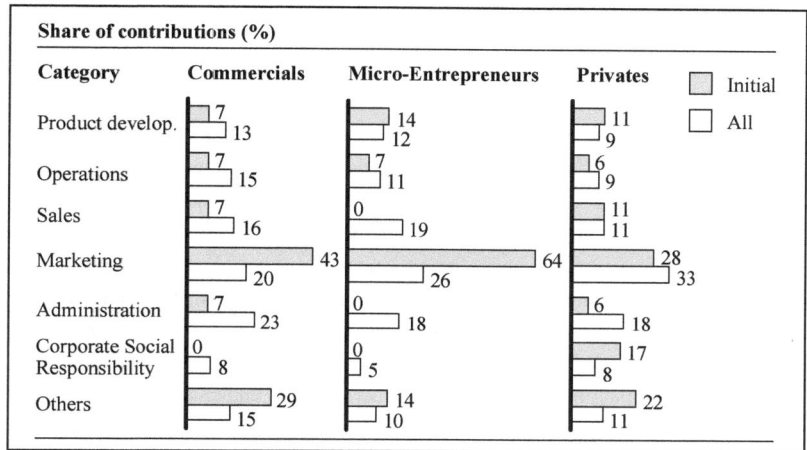

Figure 23: Initial contributions of 'Mature sample'

The initial contributions unveil a different behavior for each of the three community member groups. For the four value chain categories the distribution of the Privates' contributions show almost no differences between initial and overall. On the contrary the Commercials join the community outside of their specialized categories mostly with Marketing threads. The Micro-entrepreneurs join the community extensively focused on Marketing threads. Even though a slight focus on threads from this category is identified in the first part of the empirical study, the magnitude is astonishing. This peculiar joining pattern will be further addressed in the interviews. Considering the supporting value activities, all member groups show almost the same contribution behavior. New community members notably join the community in the category Others and avoid Administration threads. This behavior could have been easily predicted in advance as von Krogh et al. (2003) already identified the tendency to contribute to easy-to solve-problems when joining. Joining the community by introducing themselves to the community or commenting on an interesting link is clearly the simpler way for all members than contributing initially to finance or personnel threads without knowing the community comprehensively. The high share of Privates joining with Corporate Social Responsibility threads might be explained by the public image of 'Premium' as being highly concerned about sustainability.

6.3 Social network analysis

The social network analysis is widely applied in research of OIC, e.g. to identify the most influential community members (Toral et al. 2009) or knowledge brokers (Sowe et al. 2006). "The social network approach is grounded in the intuitive notion that the patterning of social ties in which actors are embedded has important consequences for those actors" (Freeman 2004, p. 2). A social network is formally defined as a set of network members, usually called nodes, which are related to each other in at least one way (Wasserman and Faust 2007). The nodes are most commonly individuals or organizations but can also represent web pages, journal articles, positions or departments within organizations, neighborhoods or even countries (Marin and Wellmann 2011). The relations between the single nodes are usually called ties, which are differentiated into directed vs. undirected and weighted vs. unweighted. Directed ties point out a path from one node to another, whereas undirected ties only indicate the relation between two nodes with no specific direction. Weighted ties carry a certain value, representing for example the level of resource transfer, whereas unweighted ties just showing a relation between two nodes without any statement about the intensity (Marin and Wellmann 2011).

One major advantage of the social network analysis is the variety of analysis levels. Most analyses start with observing the dyadic data, meaning the links between each pair of nodes rather than of each node (Borgatti and Foster 2003). Besides the dyadic level of analysis, also the actor or even the complete network level can be the entity to be studied. Measuring various centralities of each node in networks is a well-established method to gain insights on the actor-level. According to Freeman (1979), centralities can be divided into two generic classes, point and graph centralities. Point centralities consider the degrees directly allocated to one node (e.g. number of relations to other nodes) whereas graph centralities additionally consider the paths through a specific node and therefore its location within the network. The variety of centralities is diverse and therefore only centralities applied in the remainder of this thesis are explained in detail. All centralities requiring directed links are excluded automatically based on the fact that social networks drawn from co-citation data only consist of undirected ties.

Referring to the purpose of conducting this SNA, to analyze the encapsulated position of the Micro-entrepreneurs, total degree, betweenness and eigenvector centrality are chosen as measurements (Wasserman and Faust 2007):

- *Total degree centrality* measures the direct links of a node. Individuals with a high total degree centrality therefore have many connections to others in the network.

- *Betweenness centrality* represents the probability that a node is contained in the shortest paths between all different pairs of nodes. A high value in this centrality is tantamount to a brokering or gatekeeper function of the respective node.

- *Eigenvector centrality* measures the centrality of the direct neighbors of a node. Therefore, a high eigenvector centrality implicates that the node is embedded within a strong clique in which the nodes are firmly connected to each other.

In order to visualize the social network, the OSS ORA is used. The result of the co-citation analysis detailed in the fourth chapter, the CoCit-Score matrix with 75 community members serves as the data source for the network. Therefore the four different member groups Coordinators, Commercials, Micro-entrepreneurs and Privates are included in this visualization. Referring to the previously defined general structure of SNA, the members are visualized by nodes, which vary in color depending on the member group and the CoCit-Scores between the individual members by lines. For reasons of clarity thresholds are commonly used, meaning that a connection between different individuals is only displayed if the CoCit-Score between them exceeds the threshold value:

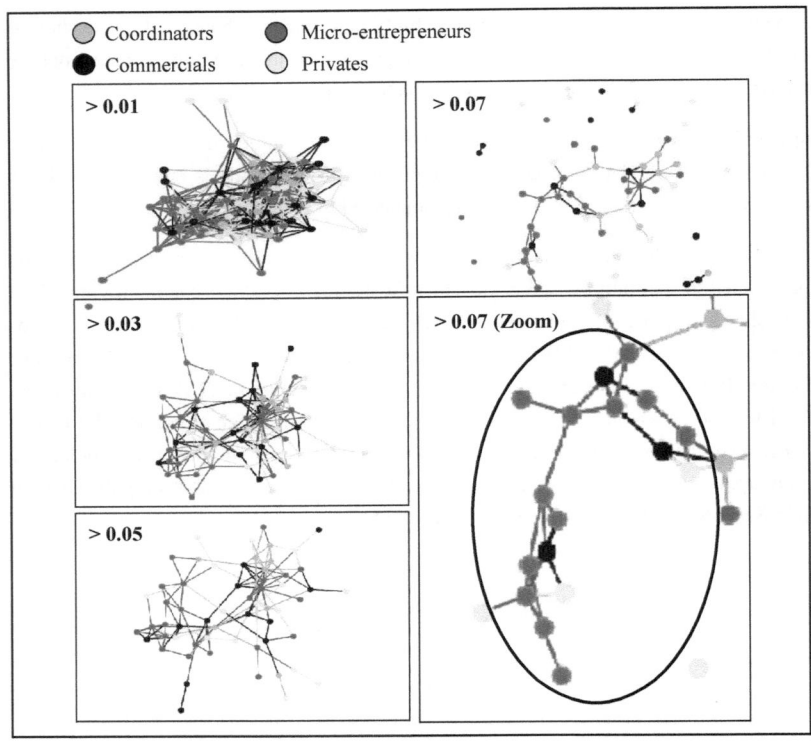

Figure 24: Visualization of the social network using different threshold values

The high density of the network with a threshold value of 0.01 hinders the conjecturing of a clear structure. With an increasing threshold a two-centered network is recognizable. Referring to the visualization with a threshold of 0.07 a right center with several Coordinators, Micro-entrepreneurs and some Commercials as well as Privates can be identified. The figure additionally shows a second center attached to the right center by only two Coordinators. By zooming into this second center a network of twelve Micro-entrepreneurs tying three Commercials and four Privates is unveiled. This number of Micro-entrepreneurs represents almost 50% of all Micro-entrepreneurs included in this analysis[56]. This encapsulated formation shows no tight communication ties to the Coordinators, making them less controllable for the community management. Additionally it offers the unique opportunity to them to significantly drive the community in a chosen direction. In order to

---

[56] 27 Micro-entrepreneurs are included in the used CoCit-Score matrix

investigate this behavior in detail on an actor-level, the following table shows different mean centralities of the three community member groups:

Mean centralities (10^-2)	Total Degree	Closeness	Betweenness	Eigenvector
Commercials	0.81	47700	1.49	31.7
Micro-entrepreneurs	0.92	47700	**1.33**	**36.5**
Privates	0.7	48100	1.43	28.2

Table 32: Centralities for the different member groups from SNA

Compared to the Commercials and Privates, the Micro-entrepreneurs have a high total degree centrality and therefore are linked to many others in the network. This can still represent a high likelihood of forming separate cliques as these many linkages could be mostly to members of their own peer group. In fact, the values of the other centrality clearly point out this isolation of the Micro-entrepreneurs. Their relatively low value for the betweenness centrality shows their low brokering and gate-keeping function. They do not serve as a connecting link between different groups. Their high value for the eigenvector centrality additionally supports the clique theory. As mentioned previously, if a node is connected to others, which are themselves highly connected to each other, ifthas a high eigenvector centrality value. Therefore, members of strong cliques show a high value in this centrality.

6.4 Interview appraisal

Interviews are a qualitative method to collect data by asking informants questions for example about their behavior, experiences or feelings (Shah and Corley 2006). Nevertheless, "interviewing is not merely the neutral exchange of asking questions and getting answers. Two (or more) people are involved in this process and their exchanges lead to the creation of a collaborative effort called the interview" (Denzin and Lincoln 2005, p. 696). Especially when using archival data such as pre-existing documents or email exchanges, interviews and observations proved to be very helpful for understanding the phenomenon and the context in which it occurs (Shah and Corley 2006). The most commonly used differentiation criterion for interviews is the classification into one of the following three types: structured, semi-structured, and in-depth interviews (Britten 1995). The structured interview consisting of a set of pre-determined questions is the most formal and the interview progress is standardized. On the contrary, the semi-structured interview, which is characterized by a rather loose structure with open-ended questions. The interviewer frequently diverges from the defined area to explore an idea in more detail. In-depth interviews are usually even less structured and additionally only cover a small number of issues but with a high detail level. The progress strongly depends on the answers given by the interviewees. Any combination of these three styles can occur, especially while progressing. The first half of the interview for example can be conducted with standardized questions, whereas the second half can be fully unstructured giving the interviewer the possibility to pursue any issue of interest (Patton 1987). Patton (1987) further lists six different types of questions: Behavior or experience, knowledge, sensory, feeling, opinion or belief, background or demographic.

Considering the purpose and characteristics of the interviews to be conducted, one might consider choosing in-depth interviews. The number of issues to be discussed is limited to member specialization topics, although these range from drivers to the individual and even the community level. Nevertheless, the foreknowledge gained from the quantitative analysis and the specificity of the questions regarding the individual member behavior indicate that semi-structured interviews with open-ended questions are the proper approach. For selecting the appropriate interviewees from the 103 community members, the purpose for conducting the interviews and the results from the qualitative analyses need to be considered carefully. In terms of specialization, explanatory elements for the differences in the specialization degrees between the three groups and the misrouted joining pattern of the Commercials are required. The other issue requesting additional explanation is the extraordinary behavior of the Micro-entrepreneurs. They contribute significantly more to product-related (value-chain) categories than the Privates and tend to form cliques reducing their knowledge exchange with the community management. Considering these, the interviewees from the Commercial group should have joined the community outside of their specialized category Operations & Sales

122

and their Gini index should be rather low. The Micro-entrepreneurs to be interviewed should also have quite a low Gini index, a relatively high share of contributions to product-related topics and a combination of a rather low betweenness centrality and a high eigenvector centrality. The Privates selected for this process should rather have a high degree of specialization, represented by a high Gini coefficient, and compared to the Micro-entrepreneurs a lower share of contributions to product-related topics. The following table shows the community members finally selected for the interviews:

No.	Member	Group	Gini index	Joining	Centralities		Contribution focus	
					Between	Eigenv.	Product	Comm.
1	5	MicroEntrepr	0.45		0.2	48.2	0.82	0.18
2	19	MicroEntrepr	0.19		2.9	77.2	0.59	0.41
3	20	Privates	0.51				0.62	0.38
4	23	Commercials	0.25	Marketing			0.49	0.51
5	30	Commercials	0.45	Marketing			0.56	0.44
6	47	MicroEntrepr	0.30		2.1	99.6	0.62	0.38
7	63	Privates	0.59				0.46	0.54

Table 33: List of community members to be interviewed

Altogether seven semi-structured telephone interviews were conducted from September to October 2012. All interviews were fully transcribed based on the audio records taken during the interviews. The interview guide consisted of two types of prepared questions. The first one included general questions regarding the motives of the members and the behavior observed for all three member groups. The second part was altered by including questions specifically for one of the three member groups. As common for semi-structured interviews, the prepared questions only functioned as rather loose guideline. The interviewer frequently diverged from these questions to deepen the understanding or to gain additional insights exceeding the defined area. Overall, five main findings emerged from the interviews providing valuable explanations for the observed behavior:

Entrepreneurial opportunities are the main motive for Micro-entrepreneurs to engage in the community.

All interviewed members from the group of Micro-entrepreneurs state the urge for building up new business as a major participation motive. They cherish the various degrees of freedom given by the community and its sponsor when establishing and expanding the selling of

'Premium' products:

> *"The reason for participating was to expand and carry out Premium." (2)*

> *"My actions are independent of others. (...) I just go out to the bars and restaurants, present Premium and just talk about Premium and then finally it is sold at the place. My decisions in this process are completely independent." (1)*

The private and commercial community members mention several motives for joining the community, such as the personal ties to community members or for general business purposes:

> *"I have known Uwe[57] for some time and decided to join 'Premium'" (7)*

> *"I was looking for an alternative to other beverages. (..) For me it is business" (5)*

Micro-entrepreneurs actively seek new knowledge instead of relying solely on knowledge already available.

Most of the community members utilize so-called 'local information', knowledge already in the possession of the member, when contributing to the community.

> *"I am not actively looking for knowledge. In moments in which I have the feeling that my contributions are valuable, are the moments in which I use expertise that I have acquired in my day-to-day business." (3)*

This behavioral pattern of user innovators is well-known (Lüthje et al. 2005). On the contrary, the Micro-entrepreneurs actively seek new knowledge by utilizing the internet or surrounding individuals. These members also rely on knowledge already in their possession. Nevertheless, in order to contribute to a greater variety of categories, they invest additional time.

> *"I would say to maybe 25 percent of all discussions I seek background information, especially technical knowledge or special knowledge." (2)*

> *„There are topics, I decide intuitively but there are also other topics, when I look for additional knowledge or ask other people around me (...)." (1)*

---

[57] The central organizer of 'Premium' and the community

Entrepreneurial attitude or a natural low cost position inhibits the lower degree of specialization.

Commercial members feature a relatively low cost position compared to all other community members, as they are more familiar with the consumer product business. Even when discussing complex administrative or operational context they can rely on knowledge already available or their expertise. As a result, they are able to spread their contributions to different categories:

> *"I have, sounds strange, but I have an overall overview. (...) I have the luxury to know all about the sensitivities of gastronomies, dealers and distributors." (5)*

The Micro-entrepreneurs cannot rely on this low cost position. Nevertheless, they try to contribute to the entire spectrum of categories to improve Premium and its products. Their previously mentioned peculiarity to contribute beyond their existing knowledge level certainly supports this behavior:

> *"I am in an area, where there are not so many established structures (...). I feel responsible for all kind of topics, because almost every topic is getting important at some point in time." (2)*

On the other hand the private members seem to stay in their own comfort zone when contributing to the community resulting in a relatively high specialization degree:

> *"We are so many people in the mailing-list, so I always thought that people should participate according to their skills. I decided that for me. I know, there are some topics I have never contributed, because I thought my input or opinion is worthless for the community." (3)*

> *"I participate only if I know exactly what to say" (7)*

The urgency for realization affects the contribution focus of the Micro-entrepreneurs

Confronted with their focus on product-related categories, the interviewed Micro-entrepreneurs explain this behavior with their need for tangible results. They seem to cherish contributions to Product Development, Operations or Sales as the results of their contributions directly affect the performance of the Premium business. On the other hand the success of contributions to the community-related categories is difficult to measure and only achieved with some time delay.

"I mean, contributing to these topics means to implement something and maybe it is because the diffusion is important to me. That Premium is getting more recognized and expanded." (1)

General Online Community inexperience fosters a misled joining of new community members.

The commercial community members interviewed had not engaged in any OIC project prior to joining the 'Premium' community. Their joining process was completely unstructured leading to an initial contribution opposing their specialization area or simply on easy-to-solve problems:

"Like I said, in such a setting (...) I have never participated. In the beginning, I have not had an inhibition level (...), it just started. I have to say, that today, I carefully consider where to contribute and where not." (5)

Part C: Summary and discussion

This final part of the thesis starts by synthesizing the results from both the quantitative and qualitative analyses of the empirical study. Subsequently the results are thoroughly discussed and explanations for the observed behavior of the community members are identified. After defining the implications of this study for research and management, the thesis closes with a brief and precise conclusion including the major outcomes from investigating the community member specialization phenomenon.

7 Summary of results

The following table summarizes the results from the empirical study. They comprise findings for a member's contribution focus, the specialization degree, the influence of the specialization on the activity level and the socialization within the group:

	++ strong ↑ -- weak	Community member group		
		Commercials	Micro-entrepreneurs	Privates
Contribution focus	Product	-	+	-
	Community	+	-	+
	Familiarity	++	+	++
	Dynamic aspect	No	Yes	No
Spe.	Degree of specialization	--	-	++
Activity level	Influenced by importance of opposed category	No	No	Yes
	Influenced by importance of specialized category	No	Yes	No
Collaboration	Controllability by community core members	++	-	++
	Likelihood of clique formation with specialized peers	-	++	-

Figure 25: Summary of results from the empirical study

As discussed previously the various community members differ in more than just their familiarity. Their familiarity with the community outcome, product-relation in this study, represents the main driver for focusing their contributions. However, this study additionally unveils many differences between the community member groups when thoroughly analyzing aspects around member specialization in OIC and their implications. Considering the results for the contribution focus of each group, they all show the predicted focus on topics close to their familiarity with the community outcome, although the group of Micro-entrepreneurs shows a slightly weaker focus. These community members additionally focus more intensively on product-related topics, represented by the value chain categories, than members from other groups. Furthermore, they are the only group that features a considerable dynamic drive in their contribution focus. Members of this group focus solely on their specialized categories when joining the group and start to generalize over time. The commercial and private members do change their contribution focus after joining the community, but this change is either rather obvious, joining the community with easy-to-solve problems[58], or simply by coincidence, inexperienced members joining an OIC for the first time[59].

Whereas the commercial and private members show similar behavioral patterns regarding their contribution focus, these groups differ in terms of their specialization degree and the influence of the community focus on their activity level. The Commercials have a lower degree of specialization in two different ways. First, they do not contribute so often to the same thread and second, they allocate their contributions more equally to all categories. In contrast, the members from the private group contribute to fewer categories and if involved in a thread they discuss more intensively resulting in a higher degree of specialization. When the community focus shifts towards topics opposed to the specialization area of the Privates, these community members reduce their activity level significantly. This activity level change is tremendously amplified if the communication volume within the community rises. On the contrary, with a changing community focus the Commercials do not adjust their activity level significantly. The specialization degree of the Micro-entrepreneurs is slightly higher compared to the Commercials but considerably lower compared to the Privates.

The activity of this group is not influenced by the importance of categories opposed to their specialization area at all. However, this group is the only one that significantly aligns its activity level to the importance of its specialized category.

---

[58] Especially in the category Others to introduce themselves or share interesting links. Research scholars, e.g. von Krogh et al. 2003, also observe this community joining behavior.
[59] As quoted in the interviews by the group of commercial community members.

Regarding the last cluster of results targeting the collaboration between the different community members, Commercials and Privates show exactly the same behavior that is predicted in the research setting chapter. In the first part of the empirical study, a focus on discussion threads with a high likelihood of peers participating is identified. Nevertheless, individuals from both community member groups clearly show the highest collaboration with the core members of the community, namely the Coordinators. Their position within the community network is characterized neither by a high likelihood of clique formation nor by a limited knowledge brokering activity. Therefore, members from the private and the commercial groups are easily controllable by the community management and their knowledge can be efficiently transferred into the community on a permanent basis. On the other hand the Micro-entrepreneurs show no significantly higher collaboration with the core members of the community compared to their own peer-group. Similar to the core members of the community, these individuals have many links to other community members as seen in the SNA. However, their central position does not result in a high level of knowledge brokering activities but rather in forming strong relationships with their own peers. This strong tendency to form cliques with likeminded reduces the influence of the community core members on the group of Micro-entrepreneurs making their behavior less controllable.

8 Discussion

In this chapter, the outcomes of the empirical study are discussed comprehensively to identify additional explanatory elements for the behavior of community members and to assess the impact of the findings in various dimensions. Firstly, drivers for the behavioral anomalies of community members not predicted when deriving the research hypotheses are identified. Secondly, the contribution of the empirical study and the chosen methodological approach for the research of OIC are detailed. Thirdly, limitations of the conducted study conditioned by the chosen research method and case are outlined and possible future research directions for the phenomenon are sketched. Lastly, clear guidance for the management of OIC is given based on the enhanced understanding of community behavior resulting from this thesis.

8.1 Discussion of empirical findings

Community members from the commercial and private group mainly act in accordance with the theories of individual behavior and empirical findings others. Based on the utility-maximization paradigm (Besanko and Braeutigam 2007) they focus their contributions on topics close to their familiarity with the community outcome as already observed in a different context by other research scholars (Nambisan and Baron 2010). This behavior is driven by their hopes for recognition and reciprocity by peers, and their applying of knowledge already in their possession (Lüthje et al. 2005). Even though members of these two groups collaborate intensively with their own peers, they hold their strongest communication ties with the group of core members, respectively the community management, on a very high significance level. This result is completely in line with other empirical studies about core members and members with dedicated roles (Dahlander and Wallin 2006; Dahlander and O' Mahony 2010; Lee 2011). However, some findings from the empirical study unveil certain differences between individuals from these two groups that are not predicted in the first part of this thesis.

Firstly, the Commercials do not align their contribution activity level to an alteration in the community focus. This can be primarily explained by their distinctive low-cost position for all topics discussed in the community as mentioned in the expert interviews. Without investing much time, they can easily read, understand and discuss all kinds of threads, even the ones from non-familiar[60] categories. On the contrary, the activity level of community members

---

[60] Even though e.g. Marketing topics are not a specialization area of the Commercials, their costs for contributing to this kind of topics is rather low as no specific expertise is required.

from the private group significantly drops with an increased importance of topics opposed to their specialization field. This may result from their distinctively weaker cost position on these topics. Screening, understanding and successfully contributing to these non-familiar topics[61] are cost-intensive and, especially in combination with an increased monthly communication volume, they reduce their activity level compared to other member groups. However, both groups, the Commercials and Privates, have in common that they do not increase their activity level when the focus of the community discussion shifts towards topics highly beneficial for them. The major insight from these findings is that costs for, and not benefit from, contributing to the community seem to be the primary driving force behind the contribution activity level of specialized community members.

The second unveiled difference between the two member groups is their different degree of specialization. Compared to the private members, the commercial members show a lower degree of specialization in two ways. On the one hand, they allocate their contributions more equally to the different categories discussed and on the other hand, they contribute to the same thread considerably less frequently. Their distinctive lower-cost position compared to the Privates may be the reason for their more heterogeneous contribution distribution. As outlined previously, they feature no major cost disadvantage in any of the discussed threads and therefore can more easily contribute to a greater variety of threads. The higher contributions per thread ratio of the private members may be caused by their primarily socially-driven participation. They enjoy discussions with other community members, sticking to their opinion, more intensively than the commercial members. The lower degree of specialization of the Commercials, necessarily leading to a higher degree of generalization, seems to be surprising as they neither feature a considerably high contribution activity[62] nor are they core members of the community. Community members showing a high degree of generalization commonly feature at least one of these two characteristics (von Krogh et al. 2003). However, an explanation for the high generalization degree of core community members may be exactly their relatively low-cost position compared to other members. As a result of their intense involvement and deep anchorage within the community, they are more familiar with the different topics and members. They can contribute to threads from different categories less cost-intensively as they can rely on previous experience and knowledge already in their possession. Additionally their barrier to contribute to certain threads is lowered as they already know most of the participating members.

---

[61] Especially threads from the category Operations and partially Product Development require deep expertise in the respective field for understanding and successfully contributing. Most of the members from the private group do not possess this expertise.

[62] The group of commercial community members is the one with the lowest mean contribution activity (29 mails, compared to approx. 42 mails for the Micro-entrepreneurs and Privates)

In contrast to the commercial and private members, the group of Micro-entrepreneurs shows a noticeably different behavior in the majority of the behavior categories. Similar to the others, they focus their contributions on topics close to their familiarity with the community outcome. However, they substantially differ in aligning their activity level and their collaboration with other community members. These are surprising findings as members from this group are basically Privates only enriched by their entrepreneurial tasks. Their individual community key figures differ only slightly from the ones private members achieve[63]. Therefore, the entrepreneurial characteristics of these community members seem to considerably drive their 'abnormal' behavior[64]. Thus, in order to explain the different contribution behavior of this group, a brief discussion of entrepreneurial characteristics in general is compulsory.

The various characteristics associated with entrepreneurs are commonly allocated to the two main theories applied in entrepreneurship research, the need for achievement and the internal locus of control (Littunen 2000). Besides these two main categories, researchers often identify a risk-taking behavior and an innovative creative style as additional categories for characterizing entrepreneurs (Caird 1993). Individuals featuring a high need for achievement are inspired by situations of high competition, where they independently solve problems and set targets, which are achieved by their own efforts (McClelland 1967). Personal characteristics often associated with a high need for achievement frequently include planning ability, self-initiative, goal-orientation, problem-solving skills and self-awareness (McClelland 1967). Referring to Rotter /1996), an individual's locus of control can be either external or internal[65]. Following the external control view, individuals relate consequences of actions to actions of others or just fate or luck. On the contrary, individuals with a rather internal locus of control see the outcome of actions as only depending on their own efforts or abilities. This belief in one's own ability to influence events is identified by several researchers as a critical characteristic that distinguishes an entrepreneur from other people, such as small business owners, e.g. (Cromie and Johns 1983; Begley and Boyd 1988). The need for control, a high grade of autonomy and independence orientation are characteristics allocated to the internal locus of control theory (Hornaday and Aboud 1971). As outlined previously, complementary to the two main theories, entrepreneurs are seen as risk-taking individuals with a rather innovative creativity. The strong correlation between risk-taking

---

[63] As outlined previously in the sixth chapter the group size, the mean activity and its standard deviation differ only slightly between Micro-entrepreneurs and Privates.

[64] The importance of the entrepreneurial task for the Micro-entrepreneurs was also stated frequently in the conducted interviews.

[65] Other authors such as Levenson distinguish the locus of control into three categories: Internal, control by others and control by chance (Levenson 1981). He argues that the control by others is at least partly influenceable by the individual. Whereas the control by luck or fate cannot be affected by the individuals themselves. However, as this thesis only focuses on the internal locus of control and the external locus of Rotten's theory includes the control by others and chance, the two-dimensional scale seems to be appropriate.

behavior and entrepreneurial orientation, already proposed by McClelland in addition to the need for achievement (McClelland 1967), explains why entrepreneurs tend to accept higher risks if challenges occur (King 1985). Their creative style of problem solving is rather innovative than adaptive, meaning that entrepreneurs tend to change the structures themselves instead of improving the existing ones (Kirton 1976).

The outlined characteristics associated with entrepreneurship help to explain the observed anomalies in the behavior of the Micro-entrepreneurs. The group of Micro-entrepreneurs focuses substantially more on threads from the value-chain categories, showing an intense focus on product-related topics. The need for achievement characterizing entrepreneurs in general might drive this observed behavior. Developing, realizing and finally bringing new products and services to the customer help to satisfy this need for achievement, presumably more than solving organizational or cultural problems of the community or firm. Especially compared to the members from the private group, they focus their contributions more intensively on topics at the beginning of the value chain. This might be due to their innovative creativeness. Instead of solving problems in a rather restricted frame, only discussing how a given product is sold or marketed, they prefer to innovate from scratch, creating entirely new products. Contrary to all other community members, Micro-entrepreneurs join the community in an extremely specialized way but reallocate their contribution focus over time. Due to their lower risk propensity, community members without any entrepreneurial characteristics may rather tend to stick to a specific contribution area, once it has been chosen. In addition, the need for achievement and their internal locus of control might drive entrepreneurs to start generalizing. Their high grade of self-initiative and their belief in their own abilities presumably direct them into other contribution areas instead of sticking to the one initially selected.

As the cost-position of the different community members is mainly based on their previous experience and available knowledge, the Micro-entrepreneurs principally need to invest the same amount as the Privates to successfully contribute to each of the categories. However, they show a considerably lower degree of specialization. The risk propensity and the need for achievement characterizing entrepreneurs might compensate for the relative weak cost position of the Micro-entrepreneurs on certain categories. Firstly, they tend to ignore the risk of reputation loss when contributing to non-familiar topics or discussion with unknown participants. Secondly, their self-initiative, goal-orientation and superior problem-solving skills enable them to contribute to topics, where they are not able to rely on knowledge already available or their experience. As a result, they tend to contribute to a greater variety of threads and therefore feature a lower degree of specialization. The Micro-entrepreneurs' activity level is positively influenced by an increased importance of their specialization area. As opposed to members from other community groups, the Micro-entrepreneurs seem to obviously recognize the possible beneficial effect of a changing community focus. The urge of entrepreneurs to actively seek opportunities and accept challenges might well cause this special behavioral pattern of the Micro-entrepreneurs. However, the behavior of these members is surprisingly not characterized by a decline in their contribution activity when the importance of non-familiar categories rises. The internal locus of control theory might deliver an appropriate explanation for this behavioral pattern. The Micro-entrepreneurs seem to ignore the external environmental changes, the unbeneficial change of the community focus, and still believe in their own ability to substantially influence the community and contribute successfully to the joint development.

According to the conducted co-citation analysis and the results from the SNA, the Micro-entrepreneurs tend to form separate cliques restricting the collaboration with the community management. Their knowledge brokering activities are limited due to their strong focus on collaborating only with members from their own group. Referring to the internal locus of control theory, the Micro-entrepreneurs cherish independency and autonomy. In contrast to other community member groups, this might lower the likelihood of the community management to get viable access to these rather closed cliques of like-minded entrepreneurial individuals. The following figure finally summarizes the observed but unpredicted behavior of the group of Micro-entrepreneurs and the assumed rationalities behind these patterns:

		Entrepreneurial characteristics			
		Need for achieve-ment	Internal locus of control	Risk-taking propensity	Innovative-ness
Observed behavior of Micro-entrepreneurs	Focus on product related topics (value-chain)	x			
	Focus on topics at the beginning of the value chain				x
	Reallocation of contribution focus after joining community	x	x	x	
	Higher degree of specialization compared to private members	x		x	
	Aligning of activity to expected benefits rather than cost position		x	x	
	Limited controllability by community management		x		
	Formation of cliques and reduced knowledge brokerage		x		

Figure 26: Micro-entrepreneurs behavior and the assumed underlying entrepreneurial characteristics

8.2 Contributions to research

This thesis provides insights for today's research in three ways. Firstly, the broader phenomenon of member behavior in OIC is conceptualized and new methodological approaches for studying this phenomenon are introduced. Secondly, the literature about member behavior and especially member specialization is considerably enriched by empirically studying behavioral patterns of different community members. Lastly, entrepreneurial characteristics are identified as an additional major explanatory element for community member behavior.

As the research field of OIC is maturing more and more, a classification into motivation, community organization and competitive dynamics (von Krogh and von Hippel 2006) seems to be outdated. Motivations of the participants and the different community organizational elements and governance rules are important elements for explaining the micro dynamics within OIC. Nevertheless, a deeper understanding of the micro dynamics within these communities, especially of the individual contribution behavior is required. Even though plenty of researchers emphasize this fact (e.g. von Korgh et al. 2003; Lakhani and von Hippel 2003), a conceptual framework synthesizing research on this phenomenon is lacking. Therefore, in this thesis an initial framework clustering community member behavior into five behavior categories is derived. This framework unveils several research gaps, especially on community joining and member specialization and might serve as a foundation for future research on member behavior. By transferring the co-citation analysis from studying bibliometric data to community communication data, the methodological toolbox for thread analysis is enriched. Particularly for undirected networks, such as mailing lists, this analysis offers the opportunity to disclose 'invisible colleges' of community members and groups by simultaneously eliminating distortions simply by a high contribution activity. Another methodological supplement is the applied Gini coefficient. This indicator usually applied for measuring income inequalities in specific countries (Kakwani 1977) has already been used for studying micro dynamics in OIC (Kuk 2006). Nevertheless, determining the specialization degree of individual community members by deploying this coefficient can be seen as an additional contribution to research.

The findings of this thesis mainly contribute to the phenomenon of member specialization in OIC. To the knowledge of the author, besides the work of Nambisan et al. (2010), this study is the only empirical work studying the contribution focus of individual community members. Besides knowledge already available as a specialization driver (von Hippel and von Krogh 2003) the close collaboration with peers are derived as additional possible driver for member specialization. Furthermore, a distinctive low cost position on a great variety of community topics is identified as a promising explanation for a high generalization degree already observed but unexplained by research scholars (von Korgh et al. 2003). Other research scholars have already emphasized the importance of core members, featuring lateral authority through obtaining specific roles (Dahlander and O' Mahony 2010) and doing coordination work (Shah 2006). This study once again proves their considerable contribution to community success, as they manage to communicate intensively with specialized members regardless of their own expertise.

By explicitly distinguishing the private community members assuming entrepreneurial tasks from the consumers only, valuable insights for OIC research are generated. So far OIC research has mainly been based on classifying community members into commercial or private (von Hippel 2007; Hars and Ou 2001). Entrepreneurship seems to stimulate innovation in online communities to some extent and vice versa (Alexy et al. 2012) and entrepreneurial characters are identified as members of OIC (Ghosh et al. 2002; Hars and Ou 2001). However, today's research completely lacks the unique characteristics shaping this community member group and its contribution behavior. Community members excited by entrepreneurship are valuable contributors to OIC as they are usually characterized by superior problem solving skills and self-initiative, for example. They go beyond their own expertise field, are focused on innovative contributions and steadily active not being afraid of unfamiliar topics. On the contrary, their pursuit for autonomy and independency due to their internal locus of control raise several challenges for OIC management.

8.3   Limitations and future research

Even having based the study on a thorough theoretical and empirical foundation and carefully considering the appropriate research methods, this study obviously has some limitations. These limitations mainly emerge from choosing a single case study approach and the applied analysis methods.

Despite the comprehensive justification of choosing a single case study, the limitation to only one mailing list necessarily jeopardizes the generalizability of the empirical findings. Nevertheless, the confidence in the transferability of the empirical findings is reasonably high for two reasons. Firstly, fundamental characteristics of the chosen case can be identified in many other OIC. In this well-established sponsored community, the only virtually linked members contribute electronically to a joint development of several designs[66] that are exploited (Raasch et al. 2009). Additionally the main level of analysis, the different

---

[66] In this case different products, services and processes

community member groups, fits perfectly into the definition of member roles (Ye et al. 2005; van Oost et al. 2008) and classifications of other researchers (von Hippel 2007; Balka et al. 2009; Hars and Ou 2001). Secondly, the empirical findings can be explained by existing theory. Previously unpredicted but observed anomalies of certain community member groups are exhaustively discussed and subsequently explained by applying additional theories, such as the need for achievement and the internal locus of control. Further limitations emerge from distinguishing the individual community members only according to their belonging to a certain peer group and simultaneously equalizing their previous experience and knowledge. This differentiation criterion is identified as one major driver for specialization and the assumption of similar experience levels and state of knowledge among peers seems to be an appropriate simplification. Nevertheless, other criteria might provide additional insights for the specialization phenomenon.

The thread-level analysis is neither unique nor inappropriate as knowledge sharing activities needs to be examined in a micro-analytical way (Kuk 2006). However, basing statistical assessments on qualitative data, such as text messages, demand great caution. Transforming unedited texts into analyzable units by coding the qualitative data requires a systematic, objective approach to prevent any distortion of the empirical findings (Krippendorff 2004; Neuendorf 2002). Therefore, in this study, the independent coders are carefully selected and the achieved inter-coder reliability complies with the highest scientific standards[67]. Nevertheless, misinterpretations of some discussion threads cannot be fully eliminated as disagreements between the coders are still observed and the inter-coder-reliability is only examined for a sufficient (Bloch, Kraemer 1989) but smaller sample. As common when using historical data, by selecting a certain data window some events might be not entirely satisfactorily recorded. On the contrary, extraordinary events distorting the empirical findings might even be included in the selected data window. Thus, in this study the entire communication data, all mails over a period of 62 months, is incorporated into the analyses. Additional confidence in the selected data window is gained by the project leader's confirming to the non-existence of any disrupting effects in the observed period[68]. Besides using qualitative data, limitations emerge from applying the co-citation analysis in this research setting. So far, this research method has been solely used in bibliometric studies of scientific communities to identify 'invisible colleges' (Gmür 2003). However, as logically reasoned in this study and with some appropriate adaptations, for example aligning the CoCit-

---

[67] The achieved Fleiss Kappa of 0.87 exceeds the almost perfect agreement threshold of 0.81 (Landis, and Koch 1977) and inter-coder-reliabilities achieved by other content analysts (Nambisan and Baron 2010; Wasko and Faraj 2005).

[68] In contrast, in the initial phase of the community in the years 2002 to 2004 the fluctuation of community members was considerably higher and the strategic direction of the community changed repeatedly.

Score threshold value, the borrowing of this method for micro-analytical studies of community communication seems to be adaptable to a great extent.

The outlined limitations point to further promising research concerning member behavior in OIC. The empirical study focuses on a sponsored OIC innovating in the tangible goods sector. An expansion to OIC from the content and software sector might unveil additional insights not investigable in this study. Furthermore, as the sponsorship of OIC projects by a corporation (Dahlander and Wallin 2006) and dedicated MOI fulfilling important tasks (Lee 2011) are favorable for the project success, widening these investigations to non-sponsored projects might be valuable. Especially the identification of community members with intense communication to members from various specialization areas is required. Although a micro-analytical approach is chosen in this study, a large-scale survey with a greater sample size among a variety of different OIC could be favorable to gain additional confidence in the empirical findings. Nevertheless, in that case some appropriate simplifications are required as a detailed thread analysis might exceed the available research resources considerably. Future research of the phenomenon should additionally use a different classification approach of the community members, e.g. content-driven vs. social-driven participants. As pointed out, the application of the co-citation analysis for investigating community communication seems promising. However, this research method should be applied in a greater variety of OIC studies to identify further helpful adaptations of this analysis for community research.

Further suggestions for future research emerge from some unpredicted but observed behavioral patterns. Particularly the dissatisfying findings for the contribution activity changes of specialized community members call for further investigations. On the one hand, future research should focus on settings in which none of the member groups features a distinctive low cost position on all thread categories discussed. On the other hand, as focusing on months with a higher number of mails considerably improved the results, communities with high communication volume should be the object of further research. Despite the detailed discussion of the behavioral anomalies of the Micro-entrepreneurs by linking them to well-established entrepreneurial theories, further research of this community member group is required: Is this member type solely attracted by the entrepreneurial task or do they also participate in OIC with limited entrepreneurial opportunities? How exactly do their entrepreneurial capabilities enable them to contribute beyond their existing knowledge? Are their contributions significantly more innovative than from other community members? Why do they tend to form separate cliques? How exactly does the clique formation of these community members affect the overall community? What mechanism can be installed to improve their knowledge exchange with the community management?

8.4 Implications for management

The discussed findings and provided explanatory approaches help the community management to improve project success in a manifold way. Several specific management measures deduced from the results directly address three main challenges for each community: attracting the required members, controlling their behavior according to the community needs and finally retaining community members. The following figure shows the proposed seven measures to be considered by managers of sponsored but also non-sponsored OIC:

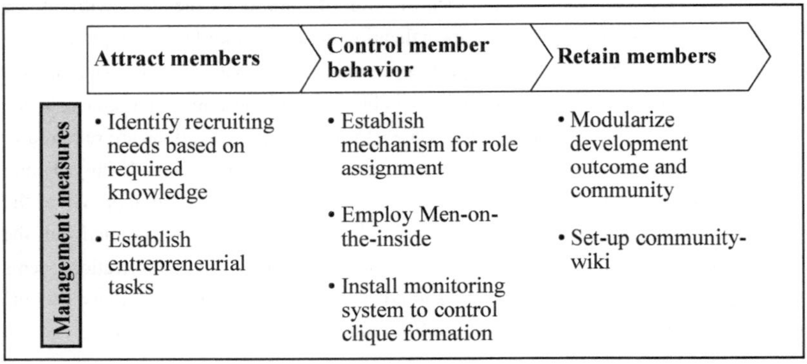

Attract members	Control member behavior	Retain members
• Identify recruiting needs based on required knowledge • Establish entrepreneurial tasks	• Establish mechanism for role assignment • Employ Men-on-the-inside • Install monitoring system to control clique formation	• Modularize development outcome and community • Set-up community-wiki

(left vertical label: Management measures)

Figure 27: Management measures along the community member lifecycle

Today firms are increasingly engaged in the sponsorships of OIC (Mahr and Lievens 2012), so the available choices for members willing to join a community is high. Therefore, the key strategic issue of OIC is to attract members (Chesbrough and Appleyard 2007). Naturally randomly attracting members seems to be the wrong strategy for each OIC. The findings of this study emphasize the need for a targeted attracting approach as most of the community members rely almost exclusively on available knowledge and previous experience when contributing. Each OIC needs to sketch a detailed knowledge heat map consisting of the knowledge required for each contribution type and the characteristics[69] of members presumably possessing this knowledge. By carefully evaluating the membership group according to these characteristics community management should be able to identify explicit recruiting needs and align their attraction measures of the OIC accordingly. As pointed out community members with entrepreneurial characteristics are valuable contributors. Consequently attracting these special individuals seems to be favorable for OIC. Although

---

[69] Possible characteristics are obviously private or commercial purpose but also for example all kinds of demographics such as location, sex, age, education

140

research on these special individuals is still in its infancy, offering entrepreneurial opportunities around the community and its outcomes looks like a promising strategy to attract these individuals.

Recommending establishing a mechanism for assigning community specific roles to individual members is hardly new to community managers as several authors already give this advice (Dahlander and O' Mahony 2010; Shah 2006). Nevertheless, the favorability of installing these individual community members for coping with specialization challenges cannot be neglected. Regardless of their own expertise, these members closely collaborate with all kinds of specialized members transferring knowledge into and/or keeping it permanently in the community. Especially in sponsored OIC, assigned MOI could also be in charge of performing these tasks. They normally grow into central positions of the network holding strong communication ties to a great variety of community members (Dahlander and Wallin 2006; Lee 2011). Even though entrepreneurial members are appreciated as contributors, their tendency to build separate cliques is challenging for the community management. However, the formation of sub-groups by highly interconnected community members restricting the control of community management quite likely occurs in various OIC settings. These 'invisible colleges' of like-minded community members can considerably jeopardize the community success. The likelihood of decomposing into two or more independent groups by forking the community is increased. This forking process is quite common for OIC but still feared as it can fuel non-productive rivalry among the different projects or simply waste valuable resources due to duplicated efforts (Cheliotis 2009). Monitoring the community network exhaustively helps the management to prevent forking or at least be well prepared for the event. In a first step, the emergence of such sub-groups should be identified at the outset. Therefore, the co-citation analysis newly introduced to OIC research paired with the social network analysis could serve as a basis for a monitoring tool designed to automatically detect these network formations. In a second step, MOI or other core members could be advised by the community management to specifically intensify their collaboration with members of this group. Tying one or more leading individuals of such groups closer to the community management is an alternative measure to presumably regain control.

An increase in participation cost considerably complicates the retaining of community members. The empirical findings of this study show a high influence of participation costs on the member's contribution activity level and degree of specialization. Members with a relatively weak cost position on a specific category of a discussed topic considerably reduce their activity if this category gains importance. These members additionally show a higher degree of specialization. Both characteristics negatively affect the retaining of community members. The likelihood of leaving the community is increased for members reacting intensively to a changing community focus and contributing only to a small portion of the discussed categories. As a result, community management should actively try to reduce the participation costs for members to improve the retaining rate. Increasing the degree of modularity and establishing a reasonable community specific knowledge base are two possible measures. Technological (de Laat 2007) and organizational (MacCormack et al. 2006) modularity is a key characteristic of OIC projects. A high degree of modularity is recommended to prevent free-riding of community members (Baldwin and Clark 2006). In addition, a high degree of modularization helps to retain members by reducing their participation costs. Members choose to participate only in favored modules. Therefore, an increased traffic in their disliked modules does not affect their participation costs, vice versa their activity level. Nevertheless, community management should carefully align the modularization degree. The highest degree of modularity might minimize the participation costs, but it obviously has some drawbacks. Firstly, modules represent sub-communities and therefore each module needs a critical mass to sustain a long-term interactive discourse (Markus 1987). Secondly, members need to join the community again and again. Before successfully participating in a module, the member needs to pass through a new joining process. Deciding in which module to participate, getting to know the participants and the culture, understanding the content discussed are cost intensive and might hinder members from extending their participation in modules chosen first. As a result, valuable contributions of members to modules outside their own comfort zone are at risk. Another measure to reduce participation costs is to set-up a community-wiki, a community specific knowledge database. Interested community members could gain additional knowledge at very low cost but valuable for successfully contributing to a great variety of different community topics. This might empower them to contribute beyond their knowledge and expertise level. The resulting lower degree of specialization helps to bond them more firmly to the community by making them less vulnerable to a changing community focus.

References

Aamodt, M. G.; Kimbrough, W. W. (1982): Effect of group heterogeneity on quality of task solutions. *Psychological Reports* 50, pp. 171–174.

Agerfalk, P. J.; Fitzgerald, B. (2008): Outsourcing to an Unknown Workforce: Exploring Opensourcing as a Global Sourcing Strategy. *MIS Quarterly* 32 (2), pp. 385–409.

Aldrich, H. E.; Herker, D. (1977): Boundary spanning roles and organization structure. *Academy of Management Review* 2 (2), pp. 217–230.

Alexy, O.; Piva, E.; Rossi-Lamastra, C. (Eds.) (2012): Citius, altius, fortius? Community-enabled bricolage and the growth of entrepreneurial ventures. *DRUID conference.* Copenhagen, June 19th - 21st.

Allen, T.J (1970): Communication networks in R & D laboratories. *R&D Management* 1 (1), pp. 14–21.

Audretsch, D. B.; Feldmann M. P. (1996): R&D Spillovers and the Geography of Innovation and Production. *The American Economic Review* 86 (3), pp. 630–640.

Backhaus, K. (2008): Multivariate Analysemethoden. Eine anwendungsorientierte Einführung. 12th ed. Berlin [u.a.]: Springer.

Bailey, E. E.; Friedlaender, A. F. (1982): Market structure and multiproduct industries. *Journal of Economic Literature* 20 (3), pp. 1024–1048.

Baker, W. E; Faulkner, R. R. (1991): Role as resource in the Hollywood film industry. *American Journal of Sociology* 97 (2), pp. 279–309.

Baldwin, C. Y. (2008): Where do transactions come from? Modularity, transactions, and the boundaries of firms. *Industrial and Corporate Change* 17 (1), pp. 155–195.

Baldwin, . Y.; Clark, K. B. (2006): The Architecture of Participation: Does Code Architecture Mitigate Free Riding in the Open Source Development Model? *Management Science* 52 (7), pp. 1116–1127.

Baldwin, C. Y.; von Hippel, E. (2011): Modeling a Paradigm Shift: From Producer Innovation to User and Open Collaborative Innovation. *Organization Science* 22 (6), pp. 1399–1417.

Balka, K.; Raasch, C.; Herstatt, C. (2009): Open source enters the world of atoms: A statistical analysis of open design. *First Monday* 14 (11).

Balka, K.; Raasch, C.; Herstatt, C. (2010): Open Source Innovation: A study of openness and community expectations. *DIME conference.* Milano, April 14th - 16th.

Bandura, A. (1977): Self-efficacy: Toward a unifying theory of behavioral change. *Psychological review* 84 (2), pp. 191–215.

Bandura, A. (1988): Organizational applications of social cognitive theory. *Australian Journal of Management* 13 (2), pp. 275–302.

Barney, J. (1991): Firm resources and sustained competitive advantage. *Journal of Management* 17 (1), pp. 99–120.

Bateman, P.; Gray, P.; Butler, B. (2006): Community commitment How Affect, Obligation and Necessity Drive Online Behavior. *International Conference on Information Systems 27.* Milwaukee.

Bechky, B. A. (2006): Gaffers, Gofers, and Grips: Role-Based Coordination in Temporary Organizations. *Organization Science* 17 (1), pp. 3–21.

Becker, G. S.; Murphy, K. M. (1992): The division of labor, coordination costs, and knowledge. *The Quarterly Journal of Economics* 107 (4), pp. 1137–1160.

Begley, T. M.; Boyd, D. P. (1988): Psychological characteristics associated with performance in entrepreneurial firms and smaller businesses. *Journal of business venturing* 2 (1), pp. 79–93.

Besanko, D.; Braeutigam, R. R. (2007): Microeconomics. An integrated approach. 3rd ed. Hoboken, N. J.: Wiley.

Blanchard, A.; Horan, T. (1998): Virtual communities and social capital. *Social science computer review* 16 (3), pp. 293–307.

Bloch, D.A.; Kraemer, H. C. (1989): 2 x 2 Kappa Coefficients: Measures of Agreement or Association. *Biometrics* 45, pp. 269–287.

Bogers, M.; Afuah, A.; Bastian, B. (2010): Users as innovators: a review, critique, and future research directions. *Journal of Management* 36 (4), pp. 857–875.

Borgatti, S. P.; Foster, P. C. (2003): The Network Paradigm in Organizational Research: A Review and Typology. *Journal of Management* 29 (6), pp. 991–1013.

Bortz, J. (2005): Statistik für Human- und Sozialwissenschaftler. 6th ed. Wien: Springer.

Britten, N. (1995): Qualitative research: Qualitative interviews in medical research. *British Medical Journal* 311 (3), pp. 251–254.

144

Brooks, C.; Ammons, J. (2003): Free Riding in Group Projects and the Effects of Timing Frequency, and Specificity of Criteria in Peer Assessments. *Journal of Education for Business* 78 (5), pp. 268–272.

Brown, J. S.; Duguid, P. (1991): Organizational Learning and Communities of Practice: Toward a Unified View of Working, Learning, and Innovation. *Organization Science* 2 (1), pp. 40–57.

Bühl, A. (2006): SPSS 14. Einführung in die moderne Datenanalyse. 10th ed. München, Boston [u.a.]: Pearson Studium.

Burt, R. S. (2005): Brokerage and closure. An introduction to social capital. Oxford, N. Y.: Oxford University Press.

Burton, R. M.; Obel, B.; Hunter, S.; Søndergaard, M.; Døjbak, D. (1998): Strategic organizational diagnosis and design: Developing theory for application. Norwell, Mass: Kluwer Academic Publisher.

Butler, B. (2001): Membership size, Communication Activity, and Sustainability: A Resource-Based Model of Online Social Structures. *Information Systems Research* 12 (4), pp. 346–362.

Caird, S. P. (1993): What do psychological tests suggest about entrepreneurs? *Journal of Managerial Psychology* 8 (6), pp. 11–20.

Cedergren, M. (2003): Open content and value creation. *First Monday* 8 (4).

Chandler, A. D. (1962): Strategy and structure. Chapters in the history of the industrial enterprise. Cambridge, Mass: MIT Press.

Chatman, J.A (1989): Improving interactional organizational research: A model of person-organization fit. *Academy of Management Review* 14 (3), pp. 333–349.

Cheliotis, G. (2009): From open source to open content: Organization, licensing and decision processes in open cultural production. *Decision Support Systems* 47 (3), pp. 229–244.

Chesbrough, H. W. (2004): Managing open innovation. *Research Technology Management* 47 (1), pp. 23–26.

Chesbrough, H. W. (2003): Open innovation: The new imperative for creating and profiting from technology. Boston, Mass: Harvard Business School Press.

Chesbrough, H. W.; Appleyard, M. M. (2007): Open innovation and strategy. *California management review* 50 (1), pp. 57–76.

Cohen, J. (1960): A Coefficient of Agreement for Nominal Scales. *Educational and Psychological Measurement* 20 (1), pp. 37–46.

Cohen, W. M.; Levinthal, D. A. (1990): Absorptive Capacity: A New Perspective on Learning and Innovation. *Administrative Science Quarterly* 35 (1), pp. 128–152.

Cromie, S.; Johns, S. (1983): Irish Entrepreneurs: Some Personal Characteristics. *Journal of Occupational Behaviour* 4 (4), pp. 318–324.

Cronin, B. (2001): Bibliometrics and beyond: Some thoughts on web-based citation analysis. *Journal of Information Science* 27 (1), pp. 1–7.

Crowston, K.; Howison, J. (2005): The social structure of open source software development teams. *First Monday* 10 (2).

Curien, N.; Fauchart; E.; Laffond, G.; Moreaus, F. (2006): Online Consumer Communities: Escaping the Tragedy of the Digital Commons. In Brousseau E., Curien N. (Eds.): Internet and digital economics. Cambridge, Mass: Cambridge University Press, pp. 201–219.

Cushway, B.; Lodge, D. (1999): Organizational Behaviour and Design. 2nd ed.: Kogan Page Publishers.

Daft, R. L.; Lewin, A. Y. (1993): Where are the theories for the "new" organizational forms? An editorial essay. *Organization Science* 4 (4), pp. 1–6.

Dahlander, L.; O' Mahony, S. (2010): Progressing to the Center: Coordinating Project Work. *Organization Science* 22 (4), pp. 961–979.

Dahlander, L.; Wallin, M. (2006): A man on the inside: Unlocking communities as complementary assets. *Research Policy* 35 (8), pp. 1243–1259.

Dahlsrud, A. (2008): How Corporate Social Responsibility is Defined: An Analysis of 37 Definitions. *Corporate Social Responsibility and Environmental Management* 15 (1), pp. 1–13.

Davis, T. R. V.; Luthans, F. (1980): A Social Learning Approach to Organizational Behavior. *Academy of Management Review* 5 (2), pp. 281–290.

De Jong, J. P. J.; von Hippel, E. (2009): Transfers of user process innovations to process equipment producers: A study of Dutch high-tech firms. *Research policy* 38 (7), pp. 1181–1191.

De Laat, P. B. (2007): Governance of open source software: state of the art. *Journal of Management and Governance* 11 (2), pp. 165–177.

Denzin, N. K.; Lincoln, Y. S. (2005): The SAGE handbook of qualitative research. 3rd ed. Thousand Oaks: Sage Publications.

Dubé, L.; Bourhis, A.; Jacob, R. (2006): Towards a typology of virtual communities of practice. *Interdisciplinary Journal of Information, Knowledge, and Management* 1 (1), pp. 69–93.

Dul, J.; Hak, T. (2008): Case Study Methodology in Business Research. 1st ed. Amsterdam: Butterworth-Heinemann/Elsevier.

Eisenhardt, K. M. (1989): Building Theories from Case Study Research. *Academy of Management Review* 14 (4), pp. 532–550.

Elo, S.; Kyngäs, H. (2008): The Qualitative Content Analysis Process. *Journal of Advanced Nursing* 62 (1), pp. 107–115.

Ernst, H. (2001): Patent applications and subsequent changes of performance: Evidence from time-series cross-section analyses on the firm level. *Research Policy* 30 (1), pp. 143–157.

Fleiss, J. L. (1971): Measuring Nominal Scale Agreement among Many Raters. *Psychological Bulletin* 76 (5), pp. 378–382.

Fleming, L.; Waguespack, D. M. (2007): Brokerage, boundary spanning, and leadership in open innovation communities. *Organization Science* 18 (2), pp. 165–180.

Foss, N. J. (2002): New Organizational Forms-Critical Perspectives. *International Journal of the Economics of Business* 9 (1), pp. 1–8.

Franke, N.; Shah, S. K. (2003): How communities support innovative activities: An Exploration of Assistance and Sharing among End-Users. *Research Policy* 32 (1), pp. 157–178.

Freeman, L. C. (1979): Centrality in social networks conceptual clarification. *Social networks* 1 (3), pp. 215–239.

Freeman, L. C. (2004): The development of social network analysis. A study in the sociology of science. Vancouver, BC: Empirical Press.

Füller, J. (2006): Why consumers engage in virtual new product developments inititated by producers. *Advances in Consumer Research* 33, pp. 639–646.

Füller, J.; Jawecki, G.; Muhlbacher, H. (2007): Innovation creation by online basketball communities. *Journal of Business Research* 60 (1), pp. 60–71.

Furman, J. L.; Porter, M. E.; Stern, S. (2002): The determinants of national innovative capacity. *Research policy* 31 (6), pp. 899–933.

Galivan, M. J. (2001): Striking a balance between trust and control in a virtual organization: A content analysis of open source software case studies. *Information Systems Journal* 11, pp. 277–304.

Garud, R.; Kumaraswamy, A. (1995): Technological and organizational designs for realizing economies of substitution. *Strategic Management Journal* 16 (1), pp. 93–109.

Gastwirth, J. L. (1972): The Estimation of the Lorenz Curve and Gini Index. *The review of economics and statistics* 54 (3), pp. 306–316.

Gemünden, H. G. (1999): Promotoren–Schlüsselpersonen für Entwicklung und Marketing innovativer Industriegüter. In J. Hauschildt, H. G. Gemünden (Eds.): Promotoren: Champions der Innovation. Wiesbaden: Gabler, pp. 43–64.

Gemünden, H. G.; Walter, A. (1999): Beziehungspromotoren–Schlüsselpersonen für zwischenbetriebliche Innovationsprozesse. In J. Hauschildt, H. G. Gemünden (Eds.): Promotoren: Champions der Innovation, vol. 2. Wiesbaden: Gabler, pp. 113–132.

Ghosh, R.; Glott, R.; Kreiger, B.; Robles-Martinez, G. (2002): The Free/ Libre/ Open Source Software Developers Survey. Available online at http://www.math.unipd.it/~bellio/.

Gmür, M. (2003): Co-citation analysis and the search for invisible colleges: A methodological evaluation. *Scientometrics* 57 (1), pp. 27–57.

Grant, R. M. (1996): Toward a knowledge-based theory of the firm. *Strategic Management Journal* 17 (Winter Special Issue), pp. 109–122.

Haefliger, S.; von Krogh, G.; Spaeth, S. (2008): Code Reuse in Open Source Software. *Management Science* 54 (1), pp. 180–193.

Hara, N.; Hew, Khe F. (2007): Knowledge-sharing in an online community of health-care professionals. *Information Technology & People* 20 (3), pp. 235–261.

Harch, B. D.; Correll, R. L. Meech W.; Kirkby, C. A.; Pankhurst, C. E. (1997): Using the Gini coefficient with BIOLOG substrate utilisation data to provide an alternative quantitaive measure for comparing bacterial soil communities. *Journal of Microbiological Methods* 30, pp. 91–101.

Harhoff, D.; Henkel, J.; von Hippel, E. (2003): Profiting from voluntary information spillovers: How users benefit by freely revealing their innovations. *Research policy* 32 (10), pp. 1753–1769.

Harrell, F. E. (2002): Regression modeling strategies. With applications to linear models, logistic regression, and survival analysis. 2nd ed. New York [u.a.]: Springer.

Harrigan, J. (1997): Technology, Factor Supplies, and International Specialization: Estimating the Neoclassical Model. *The American Economic Review* 87 (4), pp. 475–494.

Hars, A.; Ou, S. (Eds.) (2001): Working for Free? - Motivations of Participating in Open Source Projects. 34th Hawaii International Conference on System Sciences, Jan 3rd - 6th.

Hauschildt, J.; Chakrabarti, A. K. (1999): Arbeitsteilung im Innovationsmanagement. In J. Hauschildt, H. G. Gemünden (Eds.): Promotoren: Champions der Innovation, vol. 2. Wiesbaden: Gabler, pp. 67–88.

Hauschildt, J.; Kirchmann, E. (1999): Zur Existenz und Effizienz von Prozesspromotoren. In J. Hauschildt, H. G. Gemünden (Eds.): Promotoren: Champions der Innovation, vol. 2. Wiesbaden: Gabler, pp. 88–107.

Hauschildt, J.; Schewe, G. (1999): Gatekeeper und Prozesspromotoren. In J. Hauschildt, H. G. Gemünden (Eds.): Promotoren: Champions der Innovation, vol. 2. Wiesbaden: Gabler, pp. 159–176.

Heckathorn, D. (1993): Collective Action and Group Heterogeneity: Voluntary Provision versus Selective Incentives. *American Sociological Review* 58 (3), pp. 329–359.

Hedlund, G. (1994): A model of knowledge management and the N - form corporation. *Strategic Management Journal* 15 (S2), pp. 73–90.

Helfat, C. E.; Eisenhardt, K. M. (2004): Inter - temporal economies of scope, organizational modularity, and the dynamics of diversification. *Strategic Management Journal* 25 (13), pp. 1217–1232.

Hendriks, P. (1999): Why share knowledge? The influence of ICT on the motivation for knowledge sharing. *Knowledge and Process Management* 6 (2), pp. 91–100.

Herraiz, I.; Robles. G.; Amor, J. J.; Romera, T.; Barahona, J. M. (2006): The Processes of Joining in Global Distributed Software Projects. *Global Software Development Workshop.* Shanghai, May 23rd.

Herrnstein, R. J. (1990): Rational choice theory: Necessary but not sufficient. *American Psychologist* 45 (3), pp. 356–367.

Herstatt, C.; von Hippel; E. (1992): From experience: Developing new product concepts via the lead user method: A case study in a "low-Tech" field. *Journal of Product Innovation Management;* 9 (3), pp. 213–221.

Hertel, G.; Niedner, S.; Herrmann, S. (2003): Motivation of software developers in Open Source projects: an Internet-based survey of contributors to the Linux kernel. *Research Policy* 32 (7), pp. 1159–1177.

Hoetker, G. (2006): Do modular products lead to modular organizations? *Strategic Management Journal* 27 (6), pp. 501–518.

Hollis, M. (1977): Models of man: Philosophical thoughts on social action. Cambridge, Mass: Cambridge University Press.

Hornaday, J. A.; Aboud, J. (1971): Characteristics of Successful Entrepreneurs. *Personnel psychology* 24 (2), pp. 141–153.

Houston, D. J. (2004): "Walking the Walk" of Public Service Motivation: Public Employees and Charitable Gifts of Time, Blood, and Money. *Journal of Public Administration Research and Theory* 16 (1), pp. 67–86.

Jaruzelski, B.; Dehoff, K. (2008): Beyond borders: The global innovation 1000. *Strategy + business* 53 (winter), pp. 52–69.

Jeppesen, L. B.; Frederiksen, L. (2006): Why do users contribute to firm hosted user communities. *Organization Science* 17 (1), pp. 45–63.

Jeppesen, L. B.; Laursen, K. (2009): The role of lead users in knowledge sharing. *Research Policy* 38 (10), pp. 1582–1589.

Jones, Q.; Rafaeli, S. (1999): User population and user contributions to virtual publics: A systems model. *ACM's International Conference on Supporting Group Work*. Phoenix. New York: ACM.

Jorgensen, N. (2001): Putting it all in the trunk: Incremental software development in the freeBSD Open Source Project. *Information Systems Journal* 11 (4), pp. 321–336.

Kakwani, N. C. (1977): Applications of Lorenz curves in Economic Analysis. *Econometrica* 45 (3), pp. 719–728.

Kaplinsky, R.(2000): Globalisation and Unequalisation: What Can Be Learned from Value Chain Analysis. *Journal of Development Studies* 37 (2), pp. 117–146.

Kassrjian, H. H. (1977): Content Analysis in Consumer Research. *Journal of Consumer Research* 4 (1), pp. 8–18.

King, A. S. (1985): Self-analysis and assessment of entrepreneurial potential. *Simulation & Gaming* 16 (4), pp. 399–416.

Kirton, M. (1976): Adaptors and innovators: A description and measure. *Journal of applied psychology* 61 (5), pp. 622–629.

Kollock, P. (1999): The economics of online cooperation: Gifts and public goods in cybespace. In M. A. Smith (Ed.): Communities in Cyberspace. 1st ed. London: Routledge, pp. 220–237.

Kozinets, R. V. (1999): E-tribalized marketing?: The strategic implications of virtual communities of consumption. *European Management Journal* 17 (3), pp. 252–264.

Krippendorff, K. (2004): Content analysis. An introduction to its methodology. 2nd ed. Thousand Oaks, Calif.: Sage Publications.

Kuk, G. (2006): Strategic Interaction and Knowledge Sharing in the KDE Developer Mailing List. *Management Science* 52 (7), pp. 1031–1042.

Lakhani, K. R.; von Hippel, E.(2003): How open source software works: "free" user-to-user assistance. *Research Policy* 32, pp. 923–943.

Lakhani, K. R.; Wolf, Robert G. (2005): Why Hackers Do What They Do: Understanding Motivation and Effort in Free/Open Source Software Projects. In J. Feller, B. Fitzgerald, S.A Hissam, K.R Lakhani (Eds.): Perspectives on Free and Open Source Software. Cambridge, Ma.: MIT Press, pp. 3–22.

Landis, J. R.; Koch, G. G. (1977): The Measurement of Observer Agreement for Categorial Data. *Biometrics* 33 (1), pp. 159–174.

Langlois, R. N.; Garzarelli, G. (2008): Of Hackers and Hairdressers: Modularity and the Organizational Economics of Open - source Collaboration. *Industry and Innovation* 15 (2), pp. 125–143.

Lee, G. K.; Cole, R. E. (2003): From a Firm-Based to a Community-Based Model of Knowledge Creation: The Case of the Linux Kernel Development. *Organization Science* 14 (6), pp. 633–649.

Lee, K. R. (1996): The role of user firms in the innovation of machine tools: The Japanese case. *Research policy* 25 (4), pp. 491–507.

Lee, V. (2011): How Firms Can Strategically Influence Open Source Communities. The Employment of 'Men On the Inside'. Wiesbaden: Gabler.

Lepak, D. P.; Snell, S. A. (1999): The human resource architecture: Toward a theory of human capital allocation and development. *Academy of Management Review* 24 (1), pp. 31–48.

Lerner, J.; Tirole, J. (2001): The open source movement: Key research questions. *European Economic Review* 45 (4-6), pp. 819–826.

Lerner, J.; Tirole, J. (2005): The Scope of Open Source Licensing. *Journal of Law, Economics, and Organization* 21 (1), pp. 20–56.

Levenson, H. (1981): Differentiating among internality, powerful others, and chance. *Research with the locus of control construct* 1, pp. 15–63.

Lievrouw, L. A. (1989): The Invisible College Reconsidered. *Communication Research* 16 (5), pp. 615–628.

Littunen, H. (2000): Entrepreneurship and the characteristics of the entrepreneurial personality. *International Journal of Entrepreneurial Behaviour & Research* 6 (6), pp. 295–310.

Lombard, M.; Snyder-Duch, J. R.; Bracken, C. C. (2002): Content Analysis in Mass Communication. *Human Communication Research* 28 (4), pp. 587–604.

Lovgren, R.; Racer, M. (2000): Group Dynamics in Projects: Don't Forget the Social Aspects. *Journal of Professional Issues in Engineering and Education* 126 (4), pp. 156–165.

Lüthje, C. (2004): Characteristics of innovating users in a consumer goods field. *Technovation* 24 (9), pp. 683–695.

Lüthje, C.; Herstatt, C.; von Hippel, E. (2005): User-innovators and "local" information: The case of mountain biking. *Research policy* 34 (6), pp. 951–965.

MacCormack, A.; Baldwin, C. Y.; Rusnak, J. (2012): Exploring the duality between product and organizational architectures: A test of the "mirroring" hypothesis. *Research policy* 41 (8), pp. 1309–1324.

MacCormack, A.; Rusnak, J.; Baldwin, C. Y. (2006): Exploring the structure of complex software designs: An empirical study of open source and proprietary code. *Management Science* 52 (7), pp. 1015–1030.

Maclure, M.; Willett, W. C. (1987): Misinterpretation and Misuse of the Kappa Statistic. *Journal of Epidemilogy* 126 (2), pp. 161–169.

Mahr, D.; Lievens, A. (2012): Virtual lead user communities: Drivers of knowledge creation for innovation. *Research Policy* 41 (1), pp. 167–177.

Mann, H. B.; Whitney, D. R. (1947): On a Test of Whether one of Two Random Variables is Stochastically Larger than the Other. *The Annals of Mathematical Statistics* 18 (1), pp. 50–60.

Marin, A.; Wellmann, B.(2011): Social Network Analysis: An Introdcution. In John Scott (Ed.): The SAGE Handbook of Social Network Analysis. Los Angeles, Calif.: Sage Publications.

Markus, M. L. (1987): Toward a Critical Mass Theory of Interactive Media. *Communication Research* 14 (5), pp. 491–511.

Markus, M. L. (2007): The governance of free/open source software projects: Monolithic, multidimensional, or configurational? *Journal of Management & Governance* 11 (2), pp. 151–163.

McClelland, D. C. (1967): The achieving society. New York: Free Press.

Menges, R.; Schroeder, C.; Traub, S. (2005): Altruism, Warm Glow and the Willingness-to-Donate for Green Electricity: An Artefactual Field Experiment. *Environmental & Resource Economics* 31 (4), pp. 431–458.

Miles, J.; Shevlin, M. (2001): Applying regression & correlation. A guide for students and researchers. Thousand Oaks, Calif.: Sage Publications.

Miles, M. B.; Huberman, A. M. (1994): Qualitative data analysis. An expanded sourcebook. 2nd ed. Thousand Oaks, Calif.: Sage Publications.

Miles, R. E.; Snow, C. C.; Meyer, A. D.; Coleman Jr., H. J. (1978): Organizational strategy, structure, and process. *Academy of Management Review* 3 (3), pp. 546–562.

Moir, L. (2001): What Do We Mean by Corporate Social Responsibility? *Corporate Governance* 1 (2), pp. 15–22.

Morrison, P. D.; Roberts, J. H.; von Hippel, E. (2000): Determinants of user innovation and innovation sharing in a local market. *Management Science* 46 (12), pp. 1513–1527.

Müller-Stewens, G.; Lechner, C. (2005): Strategisches Management. Wie strategische Initiativen zum Wandel führen. 3rd ed. Stuttgart: Schäffer-Poeschel.

Nadler, D. A.; Tushman, M .L. (1999): The organization of the future: Strategic imperatives and core competencies for the 21st century. *Organizational Dynamics; Organizational Dynamics* 28 (1), pp. 45–59.

Nambisan, S.; Baron, R. A. (2010): Different Roles, Different Strokes: Organizing Virtual Customer Environments to Promote Two Types of Customer Contributions. *Organization Science* 21 (2), pp. 554–572.

Narduzzo, A.; Rossi, A. (2005): The role of modularity in free/open source software development. In Stefan Koch (Ed.): Free/open source software development. Hershey, Pa: Idea, pp. 84–102.

Neuendorf, K. A. (2002): The content analysis guidebook. Thousand Oaks, Calif.: Sage Publications.

Nonaka, I. (1994): A dynamic theory of organizational knowledge creation. *Organization Science* 5 (1), pp. 14–35.

Nonnecke, B.; Preece, J. (2000): Lurker demographics: Counting the silent. *Conference on Human Factors in Computing Systems*. The Hague, Netherlands.

Nov, O. (2007): What motivates Wikipedians? *Communications of the ACM* 50 (11), pp. 60–64.

Nov, O.; Naaman, M.; Ye, C. (2009): Analysis of participation in an online photo-sharing community: A multidimensional perspective. *Journal of the American Society for Information Science and Technology* 61 (3), pp. 555–566.

Ogawa, S. (1998): Does sticky information affect the locus of innovation? Evidence from the Japanese convenience-store industry. *Research policy* 26 (7-8), pp. 777–790.

Oh, W.; Jeon, S. (2007): Membership Herding and Network Stability in the Open Source Community: The Ising Perspective. *Management Science* 53 (7), pp. 1086–1101.

Oliveira, P.; von Hippel, E. (2011): Users as service innovators: The case of banking services. *Research policy* 40 (6), pp. 806–818.

Oliver, P.; Marwell, G.; Teixeira, R. (1985): A Theory of the Critical Mass, Interdependence, Group Heterogeneity, and the Production of Collective Action. *American Journal of Sociology* 91 (3), pp. 522–556.

O' Mahony, S.; Ferraro, F. (2004): Managing the Boundary of an 'Open' Project. *Harvard Working Paper No. 03-60*, pp. 1–50.

O' Mahony, S. (2003): Guarding the commons: How community managed software projects protect their work. *Research Policy* 32 (7), pp. 1179–1198.

Oreg, S.; Nov, O. (2008): Exploring motivations for contributing to open source initiatives: The roles of contribution context and personal values. *Computers in Human Behavior* 24 (5), pp. 2055–2073.

Ornetzeder, M.; Rohracher, H. (2006): User-led innovations and participation processes: Lessons from sustainable energy technologies. *Energy Policy* 34 (2), pp. 138–150.

Orton, J. D.; Weick, K. E. (1990): Loosely coupled systems: A reconceptualization. *Academy of Management Review* 15 (2), pp. 203–223.

Oxley, J. E.; Sampson, R. C. (2004): The scope and governance of international R&D alliances. *Strategic Management Journal* 25 (8 - 9), pp. 723–749.

Ozinga, J. (1999): Altruism. Westport, CT: Praeger Publishers.

Patton, M. Q. (1987): How to use qualitative methods in evaluation. Newbury Park, Calif.: Sage Publications.

Peter, J.; Lauf, E. (2003): Reliability in cross-national content analysis. *Journalism & Mass Communication Quarterly* 79 (4), pp. 815–832.

Pfeffer, J.; Salancik, G. R.; Leblebici, H. (1976): The Effect of Uncertainty on the Use of Social Influence in Organizational Decision Making. *Administrative Science Quarterly* 21 (2), pp. 227–245.

Piliavin, J. A.; Charng, H.-W. (1990): Altruism: A review of recent theory and research. *Annual Review of Sociology* 16, pp. 27–65.

Plambeck, E. L.; Taylor, T. A. (2005): Sell the plant? The impact of contract manufacturing on innovation, capacity, and profitability. *Management Science* 51 (1), pp. 133–150.

Porter, C. E. (2004): A Typology of Virtual Communities: A Multi - Disciplinary Foundation for Future Research. *Journal of Computer - Mediated Communication* 10 (1), pp. 5–12.

Porter, M. E. (1998): Competitive advantage. Creating and sustaining superior performance: With a new introduction. 1st ed. New York: Free Press.

Preece, J.; Nonnecke, B.; Andrews, D. (2004): The Top 5 Reasons For Lurking: Improving Community Experiences For Everyone. *Computers in Human Behavior 2* 20 (2), pp. 201–223.

Qureshi, I.; Fang, Y. (2010): Socialization in Open Source Software Projects: A Growth Mixture Modeling Approach. *Organizational Research Methods* 14 (1), pp. 208–238.

Raasch, C.; Herstatt, C.; Lock, P. (2008): The dynamics of user innovation: Drivers and impediments of innovation activities. *International Journal of Innovation Management* 12 (3), pp. 377–398.

Raasch, C.; Herstatt, C.; Balka, K. (2009): On the open design of tangible goods. *R&D Management* 39 (4), pp. 382–393.

Rafaeli, S.; Raban, D. R. (2005): Information sharing online: A research challenge. *International Journal of Knowledge and Learning* 1 (1/2), pp. 62–79.

Ramos-Rodríguez, A.-R.; Ruíz-Navarro, J. (2004): Changes in the intellectual structure of strategic management research: A bibliometric study of the Strategic Management Journal, 1980–2000. *Strategic Management Journal* 25 (10), pp. 981–1004.

Reid, S. E.; de Brentani, U. (2004): The fuzzy front end of new product development for discontinuous innovations: A theoretical model. *Journal of Product Innovation Management;* 21 (3), pp. 170–184.

Rheingold, H. (2000): The virtual community: Homesteading on the electronic frontier. Cambridge, Ma.: MIT Press (28).

Ridings, C. M.; Gefen, D.; Arinze, B. (2002): Some antecedents and effects of trust in virtual communities. *The Journal of Strategic Information Systems* 11 (3-4), pp. 271–295.

Roberts, E. B. (1988): Managing invention and innovation. *Research Technology Management* 31 (1), pp. 11–23.

Roberts, E. B.; Fusfeld, A. R. (1981): Staffing the innovative technology-based organization. *Sloan management review* 22 (3), pp. 19–34.

Roberts, J. A. (1996): Green Consumers in the 1990s: Profile and Implications for Advertising. *Journal of Business Research* 36 (3), pp. 217–231.

Roberts, J.A.; Hann, I.-H.; Slaughter, S. A. (2006): Understanding the Motivations, Participation, and Performance of Open Source Software Developers: A Longitudinal Study of the Apache Projects. *Management Science* 52 (7), pp. 984–999.

Rodriguez-Clare, A. (1996): The division of labor and economic development. *Journal of Development Economics* 49 (1), pp. 3–32.

Romanelli, E. (1991): The evolution of new organizational forms. *Annual Review of Sociology* 17, pp. 79–103.

Rosen, S. (1983): Specialization and human capital. *Journal of Labor Economics* 1 (1), pp. 43–49.

Rosenkopf, L.; Nerkar, A. (2001): Beyond local search: Boundary - spanning, exploration, and impact in the optical disk industry. *Strategic Management Journal* 22 (4), pp. 287–306.

Rost, K.; Hölzle, K.; Gemünden, H. G. (2007): Promotors or champions? Pros and cons of role specialisation for economic process. *Schmalenbach Business Review* 59, pp. 340–363.

Rotter, J. B. (1966): Generalized expectancies for internal versus external control of reinforcement. *Psychological monographs: General and applied* 80 (1), pp. 1–28.

Rowlands, I. H.; Scott, D.; Parker, P. (2003): Consumers and green electricity: Profiling potential purchasers. *Business Strategy and the Environment* 12 (1), pp. 36–48.

Ruxton, G. D. (2006): The unequal variance t-test is an underused alternative to Student's t-test and the Mann-Whitney U test. *Behavioral Ecology* 17 (4), pp. 688–690.

Ryan, R. M.; Deci E. L. (2000): Intrinsic and Extrinsic Motivations: Classic Definitions and New Directions. *Contemporary Educational Psychology* 25, pp. 54–67.

Schäffer, U.; Binder, C.; Gmür, M. (2006): Struktur und Entwicklung der Controllingforschung — Eine Zitations- und Kozitationsanalyse von Controllingbeiträgen in deutschsprachigen wissenschaftlichen Zeitschriften von 1970 bis 2003. *Zeitschrift für Betriebswirtschaft* 76 (4), pp. 395–440.

Schilling, M. A. (2000): Toward a general Modular Systems Theory and its Application to Interfirm Product Modularity. *Academy of Management Review* 25 (2), pp. 312–334.

Schilling, M. A.; Steensma, H. K. (2001): The Use of Modular Organizational Forms: An Industry-Level Analysis. *Academy of Management Journal* 44 (6), pp. 1149–1168.

Schoberth, T.; Preece, J.; Heinzl, A. (2003): Online Communities: A Longitudinal Analysis of Communication Activities. 36th *International Conference on System Sciences*. Hawaii, Jan 6th - 9th.

Schulte-Zurhausen, M. (2005): Organisation, 4., überarbeitete und erweiterte Auflage. München: Verlag Franz Vahlen.

Schweisfurth, T.; Raasch, C.; Herstatt, C. (2011): Free revealing in open innovation: A comparison of different models and their benefits for companies. *International Journal of Product Development* 13 (2), p. 95.

Sen, R.; Subramaniam, C.; Nelson, M. L. (2008): Determinants of the Choice of Open Source Software License. *Journal of Management Information Systems* 25 (3), pp. 207–240.

Shah, S. K.; Tripsas, M. (2007): The accidental entrepreneur: The emergent and collective process of user entrepreneurship. *Strategic Entrepreneurship Journal* 1 (12), pp. 123–140.

Shah, S. K. (2006): Motivation, Governance, and the Viability of Hybrid Forms in Open Source Software Development. *Management Science* 52 (7), pp. 1000–1014.

Shah, S. K.; Corley, K. G. (2006): Building Better theory by Bridging the Quantitative-Qualitative divide. *Journal of Management Studies* 43 (8), pp. 1821–1835.

Siggelkow, N. (2007): Persuasion with Case Studies. *Academy of Management Journal* 50 (1), pp. 20–24.

Simon, H. A. (1962): The architecture of complexity. *Proceedings of the American philosophical society* 106 (6), pp. 467–482.

Skiba, F.; Herstatt, C. (2009): Users as sources for radical service innovations: Opportunities from collaboration with service lead users. *International Journal of Services Technology and Management* 12 (3), pp. 317–337.

Small, H. G. (1977): A Co-citation Model of a Scientific Specialty: A Longitudinal Study of Collagen Research. *Social Studies of Science* 7 (2), pp. 139–166.

Smith, A. (1937): An Inquiry into the Nature and Causes of the Wealth of Nations. Modern Library edition. New York: Random House.

Smith, V. H.; Kehoe, M. R.; Cremer, M. E. (1995): The private provision of public goods: Altruism and voluntary giving. *Journal of Public Economics* 58 (1), pp. 107–126.

Sowe, S. K. (2008): Understanding knowledge sharing activitites in free/open source software projects: An empirical study. *Journal of Systems and Software* 81 (3), pp. 431–446.

Sowe, S.; Stamelos, I.; Angelis, L. (2006): Identifying knowledge brokers that yield software engineering knowledge in OSS projects. *Information and Software Technology* 48 (11), pp. 1025–1033.

Stinchcombe, A. L. (2000): Social Structure and Organizations. *Advances in Strategic Management* 17, pp. 229–259.

Strauss, A. L.; Corbin, J. M. (1991): Basics of qualitative research. Grounded theory procedures and techniques. 3rd ed. Newbury Park, Calif.: Sage Publications.

Teece, D. J. (1980): Economies of scope and the scope of the enterprise. *Journal of economic behavior & organization* 1 (3), pp. 223–247.

Thorn, B. K.; Connolly, T. (1987): Discretionary Data Bases: A Theory and Some Experimental Findings. *Communication Research* 14 (5), pp. 512–528.

Tietz, R.; Morrison, P. D.; Luthje, C.; Herstatt, C. (2005): The process of user-innovation: A case study in a consumer goods setting. *International Journal of Product Development* 2 (4), pp. 321–338.

Tietz, R. (2007): Virtuelle Communities als ein innovatives Instrument für Unternehmen. Eine explorative Fallstudienanalyse im Hobby- und Freizeitgüterbereich. Hamburg: Verlag Dr. Kovac.
158

Tönnies, F. (2012): Gemeinschaft und Gesellschaft. *Studien zu Gemeinschaft und Gesellschaft*, pp. 213–219.

Toral, S. L.; Martínez-Torres, M. R.; Barrero, F.; Cortés, F. (2009): An empirical study of the driving forces behind online communities. *Internet Research* 19 (4), pp. 378–392.

Townley, B. (1993): Foucault, Power/Knowledge, and its Relevance for Human Resource Management. *Academy of Management Review* 18 (3), pp. 518–545.

Turner, J. H.; Maryanski, A.; Fuchs, S. (1991): The structure of sociological theory: Wadsworth Belmont, CA.

Turner, J. H. (2003): The structure of sociological theory. 7th ed. Chicago, Ill: Dorsey Press.

Turner, R. H. (1985): Unanswered questions in the convergence between structuralist and interactionist role theories. In J. H. Helle, S. N. Eisenstadt (Eds.): Micro-sociological Theory: Perspectives on Sociological Theory. Beverly Hills, Calif.: Sage Publications, pp. 22–36.

Tushman, M. L. (1977): Special boundary roles in the innovation process. *Administrative Science Quarterly* 22 (4), pp. 587–605.

Tushman, M. L.; Scanlan, T. J. (1981): Boundary spanning individuals: Their role in information transfer and their antecedents. *Academy of Management Journal* 24 (2), pp. 289–305.

Tversky, A.; Kahneman, D. (1986): Rational choice and the framing of decisions. *Journal of business* 59 (4), pp. 251–278.

Van den Bosch, F. A. J.; Volberda, H. W.; de Boer, M. (1999): Coevolution of firm absorptive capacity and knowledge environment: Organizational forms and combinative capabilities. *Organization Science* 10 (5), pp. 551–568.

Van Oost, E.; Verhaegh, S.; Oudshoorn, N. (2008): From Innovation Community to Community Innovation: User-initiated Innovation in Wireless Leiden. *Science, Technology & Human Values* 34 (2), pp. 182–205.

Van Wendel de Joode, R., de Bruijn, J. A; van Eeten, M. (2003): Protecting the virtual commons. Self-organizing open source and free software communities and innovative intellectual property regimes. The Hague: T. M. C. Asser Press.

Volberda, H. W. (1999): Building the flexible firm: How to remain competitive. Cambridge, Mass: Oxford University Press.

Von Hippel, E. (1976): The dominant role of users in the scientific instrument innovation process. *Research policy* 5 (3), pp. 212–239.

Von Hippel, E. (1988): The sources of innovation. New York: Oxford University Press.

Von Hippel, E. (1994): "Sticky information" and the locus of problem solving: Implications for innovation. *Management Science* 40 (4), pp. 429–439.

Von Hippel, E. (2006): Democratizing innovation. Cambridge, Ma.: MIT Press.

Von Hippel, E. (2007): Horizontal innovation networks--by and for users. *Industrial and Corporate Change* 16 (2), pp. 293–315.

Von Hippel, E.; von Krogh, G. (2003): Open Source Software and the "Private-Collective" Innovation Model: Issues for Organizational Science. *Organizational Science* 14 (209-223).

Von Krogh, G.; Spaeth, S.; Lahkahni, K. (2003): Community, joining, and specialization in open source software innovation: a case study. *Research Policy* 32 (7), pp. 1217–1241.

Von Krogh, G.; von Hippel, E. (2006): The promise of research on open source software. *Management Science* 52 (7), pp. 975–983.

Wasko, M.; Faraj, S. (2000): "It is what one does": Why people participate and help others in electronic communities of practice. *Strategic Information Systems* 9 (2-3), pp. 155–173.

Wasko, M.; Faraj, S. (2005): Why should I share? Examing Social Capital and Knowledge Contribution in Electronic Networks of Practice. *MIS Quarterly* 29 (1), pp. 35–58.

Wasserman, S.; Faust, K. (2007): Social network analysis. Methods and applications. 16th ed. Cambridge: Cambridge University Press.

West, J.; O' Mahony, S. (2008): The Role of Participation Architecture in Growing Sponsored Open Source Communities. *Industry and Innovation* 15 (2), pp. 145–168.

White, H. D.; Griffith, B. C. (1981): Author Cocitation: A Literature Measure of Intellectual Structure. *Journal of the American Society of Information Science* 32 (3), pp. 163–171.

White, H. D.; McCain, K. W. (1998): Visualizing a discipline: An Author Co-citation Analysis of Information Science, 1972-1995. *Journal of the American Society of Information Science* 49 (4), pp. 327–355.

Winship, C.; Mandel, M. (1983): Roles and positions: A critique and extension of the blockmodeling approach. *Sociological methodology* 14, pp. 314–344.

Witte, E. (1977): Power and innovation: A two-center theory. *International Studies of Management & Organization* 7 (1), pp. 47–70.

Witte, E. (1999): Das Promotoren-Modell. In J. Hauschildt, H .G. Gemünden (Eds.): Promotoren: Champions der Innovation. Wiesbaden: Gabler, pp. 9–41.

Xu, J.; Gao, Y.; Christley, S.; Mades, G. (2005): A Topological analysis of the Open source Software Development Community. 38th *International Conference on System Sciences.* Hawaii, Jan 5th - 8th.

Ye, Y.; Nakakoji, K.; Yamamoto, Y.; Kishida, K. (2005): The co-evolution of systems and communities in Free and Open Source Software Development. In Stefan Koch (Ed.): Free/open source software development. Hershey, Pa: Idea, pp. 59–82.

Yin, R. K. (2009): Case study research. Design and methods. 4th ed. Los Angeles, Calif.: Sage Publications.

161

Appendix

Interview guide

Category	Question	Relevant for member group		
		Commer.	Micro-Ent.	Privates
Overall	How did you initially get in touch with the Premium community?	x	x	x
	What was the major reasoning for you to join the community?	x	x	x
	What are your motives for being an active contributor?	x	x	x
	How do you gain the knowledge required to successfully contribute to a stated problem?	x	x	x
Contribution focus	How do you usually decide if you contribute to a discussed thread or not? Any kind of decision process?	x	x	x
	What do you think is the rationale behind your contribution focus on community, respectively non-product-related topics?	x		x
	What do you think is the rationale behind your contribution focus on product-related topics?		x	
	How did you select the thread for your initial contribution?	x		
Specialization degree	Why do you tend to contribute frequently to the same thread?			x
	Why do you tend to contribute less frequently to the same thread?	x	x	
	Why do you tend to spread your contribution to a greater variety of discussion categories?	x	x	
	Why do you tend to spread your contribution to a smaller variety of discussion categories?			x
Collaboration	What might be the rationale behind your extensive collaboration with other peers?		x	